Knowledge Integration Methods for Probabilistic Knowledge-Based Systems

Knowledge-based systems and solving knowledge integrating problems have seen a great surge of research activity in recent years. *Knowledge Integration Methods for Probabilistic Knowledge-Based Systems* provides a wide snapshot of building knowledge-based systems, inconsistency measures, methods for handling consistency, and methods for integrating knowledge bases. The book also provides the mathematical background to solving problems of restoring consistency and integrating probabilistic knowledge bases in the integrating process. The research results presented in the book can be applied in decision support systems, semantic web systems, multimedia information retrieval systems, medical imaging systems, cooperative information systems, and more. This text will be useful for computer science graduates and PhD students, in addition to researchers and readers working on knowledge management and ontology interpretation.

Knowledge Integration Methods for Probabilistic Knowledge-Based Systems

Van Tham Nguyen
Ngoc Thanh Nguyen
Trong Hieu Tran

CRC Press
Taylor & Francis Group
Boca Raton London New York

CRC Press is an imprint of the
Taylor & Francis Group, an **informa** business

A CHAPMAN & HALL BOOK

First edition published 2023
by CRC Press
6000 Broken Sound Parkway NW, Suite 300, Boca Raton, FL 33487-2742

and by CRC Press
4 Park Square, Milton Park, Abingdon, Oxon, OX14 4RN

CRC Press is an imprint of Taylor & Francis Group, LLC

© 2023 Taylor & Francis Group, LLC

Library of Congress Cataloging-in-Publication Data
Names: Nguyen, Van Tham (Researcher on computational intelligence), author. | Nguyen, Ngoc Thanh (Computer scientist), author. | Tran, Trong Hieu, author.
Title: Knowledge integration methods for probabilistic knowledge-based systems / Van Tham Nguyen, Ngoc Thanh Nguyen, Trong Hieu Tran.
Description: First edition. | Boca Raton : CRC Press, 2023. | Includes bibliographical references and index. | Identifiers: LCCN 2022028082 | ISBN 9781032232188 (hardback) | ISBN 9781032233888 (paperback) | ISBN 9781003277019 (ebook)
Subjects: LCSH: Expert systems (Computer science) | Probabilistic databases.
Classification: LCC QA76.76.E95 N4995 2022 | DDC 006.3/3--dc23/eng/20220815
LC record available at https://lccn.loc.gov/2022028082

ISBN: 978-1-032-23218-8 (hbk)
ISBN: 978-1-032-23388-8 (pbk)
ISBN: 978-1-003-27701-9 (ebk)

DOI: 10.1201/9781003277019

Typeset in Latin Modern font
by KnowledgeWorks Global Ltd.

Publisher's note: This book has been prepared from camera-ready copy provided by the authors.

Contents

Preface

Inconsistency of knowledge often takes place when the knowledge is gathered from various sources. This phenomenon often occurs in social networks and generally in distributed environments. Resolving knowledge inconsistency and recovering knowledge consistency are a common subject of Knowledge Management and Knowledge Integration areas. The inconsistency of knowledge may be considered on two levels, namely *syntactic* and *semantic*. At the syntactic level, inconsistency may appear as the contradiction of logical formulas without interpretation. On the semantic level, on the other hand, inconsistency appears when referring to the same real world; these formulas have different logical values. In this sense inconsistency at the semantic level is more general than inconsistency at the syntactic level. The problem of solving knowledge inconsistency is very important because it enables to achieve the consistency of knowledge bases, which is an essential feature of the bases to function. There are many models for knowledge consistency recovering presented in the worldwide literature. In this book we deal with the probabilistic model for this aim. The need for knowledge inconsistency resolution arises in many practical applications of computer systems. This book provides a wide snapshot of some intelligent technologies for knowledge inconsistency resolution based on probabilistic approach. It completes the newest research results of the authors in the period of the last five years. For practical requirements: the research results of this research can be widely applied in decision support systems, automated e-commerce systems, semantic web systems, as well as expert systems to enhance the accuracy of disease diagnostic systems; weather forecasting systems and economic forecasts; systems to combat climate change and prevent natural disasters and epidemics; and other fields. These systems serve many aspects of social life as well as national security The authors hope that the book can be useful for graduate and Ph.D. students in Computer Science; participants of courses: knowledge engineering and collective intelligence; researchers and all readers working on knowledge management and ontology integration; and specialists of social media. For related fields of science and technology: The results of this research should provide theoretical models, evaluation results on the rationality, and computational complexity for the community to do research and development. We wish to express great gratitude to the reviewers for their valuable comments. Special thanks go to Randi Cohel from the Publisher for Computer Science and IT at Taylor and Francis Group for her kind contacts and advices in preparing this book.

Van Tham Nguyen, Ngoc Thanh Nguyen and Trong Hieu Tran

Authors

Van Tham Nguyen
Thuyloi University
Hanoi, Vietnam

Ngoc Thanh Nguyen
Wroclaw University of Science and Technology
Poland

Trong Hieu Tran
Vietnam National University
Hanoi, Vietnam

CHAPTER 1

Introduction

1.1 MOTIVATION

Inconsistency of knowledge, as defined in [1] can be understood as a situation in which a knowledge base contains some contradictions. However, the notion of inconsistency of knowledge is more general and complex than the notion of contradiction. Nguyen (2008) [1] showed that, in general, for setting an inconsistency of knowledge, the following three components are needed:

- A subject to which the inconsistency refers: If we assume that a knowledge base contains knowledge about some real world, the subject of inconsistency may be a part of the real world.

- A set of elements of knowledge related to this subject: such an element may be, for example, a formula or a relational tuple in the base.

- Definition of contradiction: In the case of logic-based bases, the contradiction is simple to be identified owing to the criterion of contradiction in the set of formulae. In the case of non-logic bases, contradiction needs to be indicated referring to the subject of inconsistency.

In the case of logic-based knowledge bases, the inconsistency of knowledge is reduced to the contradiction of a set of formulae. The definition of contradiction in this case is well-known, very strong and independent from what the subject refers to.

In the case of non-logic bases all, three components are important; without one of them it is hard to set the inconsistency.

Knowledge integration is an important task when we want to combine several knowledge-based systems into a single system. This should make them able to interact with each other. It is well-known that knowledge or data in such a system should be consistent, otherwise, it is impossible to make use of it. Thus we can understand knowledge integration as a process in which several knowledge bases are integrated (merged) to create a new knowledge base:

$$\mathcal{K}_1, \mathcal{K}_2, \ldots, \mathcal{K}_n \to \mathcal{K}$$

with the assumption that there may exist inconsistency among bases $\mathcal{K}_1, \mathcal{K}_2, \ldots, \mathcal{K}_n$. The integration process should fulfill the following two general criteria:

1) Knowledge base $\mathcal{K}$ should not contain inconsistency.

2) Knowledge base $\mathcal{K}$ should best represent the given bases $\mathcal{K}_1, \mathcal{K}_2, \ldots, \mathcal{K}_n$.

In practice, knowledge integration is a complex task because there can be many aspects of the inconsistency, and in many cases the integration problems are NP-complete [1].

This book will address these two criteria. We propose a probabilistic model for knowledge integration. We will make the criteria more concrete and work out algorithms for satisfying them.

Applications of knowledge integration take place in many fields and are numerous and varied [2].

One of the results of integrating knowledge from various sources is that the new knowledge can be discovered. However, along with this, aspect uncertainty can also occur. Thus, it is one additional problem requiring solutions.

Up to now, the following knowledge representations have been used in the knowledge integration tasks: classical logic, modal logics, probabilistic-logic, and probabilistic models.

We can, in short, characterize the existing approaches for resolving knowledge inconsistency as follows:

- Removing inconsistent logic formulas: There are two ways used. The first is based on removing the minimal set of inconsistent formulas so that the remaining formulas are consistent [3, 4, 5, 6, 7, 8, 9]. The second way is based on determining the maximal set of formulas that does not contain inconsistency [10]. The disadvantage of this approach is that the formulas are treated in quantity, not in quality (semantic aspect). One can lose important formulas.

- Believe revision: This approach is based on revising knowledge in one or more knowledge bases taking part in the integration process. It can be understood as preprocessing the knowledge by revising the logic formulas and thus the inconsistency can be resolved [11, 12]. This approach can be useful for logical representations as well as, for probabilistic models. However, this is no guarantee that revising some formulas can make them inconsistent with others and the process can exact a very high cost for achieving consistency.

For knowledge integration, we have the following main approaches:

- For classical logic and modal logic representations there have been proposed advanced and effective methods [3, 13, 14, 15, 16, 17, 18]. A knowledge base is here understood as a set of rules. The idea of the worked algorithms is most often based on fitting the rules to integrating them. For this, we can use integration functions or distance functions. Another approach is based on using consensus theory for integration, treating the formulas coming from various bases as votes that need to be reconciled [19, 20, 21, 22]. However, the disadvantage of all these methods is related to their effectiveness regarding big data, on which more and more often the knowledge bases rely upon.

- For probabilistic-logic representation, several methods have been proposed [23]. These methods use the strong feature of probabilistic theory for resolving inconsistency using logic inference mechanisms. However, the solutions to the integration problem in the probabilistic-logic environment are still limited for this approach, which must be determined by probability functions on the universe set and considered in the environment of propositional calculus and are bound by the logical formulas. Furthermore, it is also difficult to specify how new probabilities in knowledge bases can be calculated from the proposed general model. This approach will also not be feasible for problems with large input spaces. This approach just stops at solving the inconsistency of the knowledge bases.

- For non-logic probabilistic representations, which are convenient and popular today when the knowledge is often mined from data, two main approaches to solving the integration problem are:

i) Determining an appropriate joint probability distribution from the sets of probability distributions representing the input probabilistic knowledge [24, 25, 26, 27, 28, 29, 30]; and

ii) Determining the probability value of probabilistic constraints [12, 31, 32, 33]. With the approach to find a conjugated probability distribution, there are three techniques deployed: using iterative techniques [24, 25, 26]; using candidacy functions [34]; and using divergence distance functions (DDFs) [27, 28, 29, 30]. However, these approaches require that the input knowledge bases be consistent and have the same structure.

Figure 1.1 summarizes the number of works published on (`https://www.sciencedirect.com`) for the period 1997-2021 with the phrases, such as "Knowledge-base system", "Probabilistic knowledge", "Inconsistent knowledge", "Consistency knowledge", "Merging knowledge or Knowledge Integration". These phrases appear in titles, abstracts, and the list of keywords of the publication topics, such as

"Expert Systems with Applications", "Information Sciences", "Neurocomputing", "Knowledge-Based Systems" and "Procedia Computer Science".

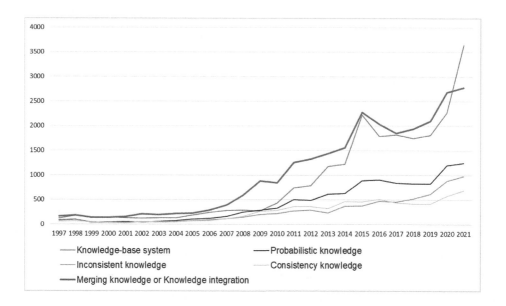

Figure 1.1 Statistics of scientific works on inconsistency-handling and integration problem solving in knowledge-based systems on ScienceDirect for the period 1997-2021.

We present our research motivation as follows.

1. To be able to solve the integration problem, first of all, ensuring the consistency of knowledge-based systems is always one of the essential requirements because if the consistency is not guaranteed, most of these systems become useless. Because of that importance, a lot of research has been interested in restoring consistency in knowledge-based systems. As noted above, the main ways to restore the consistency of a knowledge base are: removing inconsistent formulas, revising inconsistent formulas, and changing the probability (change in interval probability, change in point probability). However, these methods only stop at dealing with knowledge base inconsistencies using classical logic, possibilistic, and probability logic, or probability on a propositional language. Furthermore, there is no proposed consistency-restoring models as well as proposed algorithms to solve the inconsistency for probabilistic knowledge bases (PKBs) using sets of events. Evaluation of the reliability, correctness, as well as ensuring the basic principles of probability when changing the probability values of the probabilistic constraints in PKBs also need to be considered and proven.

2. If the consistency of the input knowledge bases is ensured then the knowledge integration process can be carried out and bring about high efficiency in terms of

integration quality as well as the satisfaction of the basic properties of the integration principle. Depending on the type of knowledge base, there will be appropriate integration methods. With the probabilistic knowledge base, the approaches to solving the problem of integrating knowledge are still facing some important problems. First, the inputs and outputs of PKBs are represented by probability distributions, where the inputs to the knowledge bases must be consistent and have the same structure. Second, there are no proposed models or algorithms to solve the problem of integrating probabilistic knowledge on sets of events. Third, with the distance function-based approach, only a few distance functions have been studied and applied and it appears that they are not so effective in calculating. Finally, the assessment of reliability, correctness, as well as ensuring the principles integrated on the probabilistic environment also need to be analyzed and proven. Therefore, there are still a lot of work to be done, such as building the theoretical frameworks, collecting the practical data as well as evaluating the experimental results.

Therefore, in order to build an integrated system of PKBs, the aim of our studies is to overcome and improve the remaining problems of consistency recovery methods for integrating the knowledge base in a probabilistic environment. For this aim, we have explored the approach to solving inconsistency and integrating knowledge bases using classical logic, possibilistic logic, and probabilistic-logic representations. Then we propose new models and algorithms to solve the problem of restoring the consistency of the PKB. Next, we study the currently developed methods of knowledge integration to find a suitable solution for the problem of integrating PKBs on event sets. Finally, based on the results of the experiments, we will analyze, compare and evaluate the quality of the obtained results, the time complexity to solve the optimization problems related to integrating PKBs.

1.2   THE OBJECTIVES OF THIS BOOK

The main subject of this book is to propose methods to restore the consistency of PKBs and methods to integrate these knowledge bases. Specifically, our objectives are as follows:

- Providing an overview report on knowledge-base inconsistency measures; methods that integrate knowledge bases using classical logic models, probabilistic-logic models, negotiation models, and probabilistic models.

- Developing a new method to represent the probabilistic knowledge base; a method of representing the inconsistency of the PKB and the general principle diagram of the system of probabilistic knowledge integration.

- Building a model to restore the consistency of the PKB. Evaluate and analyze inconsistency measures, choose the best measure that fits the probabilistic model by considering the expected properties that they need to satisfy. Proposing a new family of operators to restore uniformity for probabilistic models; investigate and develop a set of expected properties for the proposed operators. Constructing a model for the through-statement approach to the problem of restoring the consistency for PKBs. Building and evaluating the computational complexity and implementing consistency restoring algorithms.

The major trend of knowledge-based systems today is to be built to work with big data, from a variety of sources, and the knowledge source of these systems is often represented as probabilities. The integration in knowledge-based systems in which knowledge is represented in the form of logical formulas or knowledge frameworks has been studied and widely applied. However, it is difficult to apply frameworks of knowledge represented in this form to probabilistic knowledge bases. Therefore, building frameworks to integrate PKB is topical, very necessary and needs to be researched.

For practical requirements: The results of this research can be widely applied in decision support systems, automated ecommerce systems, semantic web systems, as well as in expert systems to enhance the accuracy of disease diagnostic systems; weather forecasting systems, economic forecasts; systems to combat climate change, prevent natural disasters and epidemics; and other fields. These systems serve many aspects of social life as well as national security. For related fields of science and technology: The results of this research will provide theoretical models, evaluation results on the rationality, computational complexity for the community to do research and development.

1.3   THE STRUCTURE OF THIS BOOK

The content of this book is divided into eight chapters. In the first chapter, we present the introduction, the motivation and the objectives of our research.

In the second chapter, we present the background knowledge used in this book, including an overview of knowledge representation techniques, the method of representing PKBs, and the method of representing the inconsistency in knowledge bases to solve the problem of restoring consistency in the PKB-defined Chapter 4. Next we discuss distance functions to build an integrated model of definite knowledge bases, performance and conduct experimental installation of the methods proposed in Chapter 5 and Chapter 6.

Chapter 3 presents the class of inconsistency measures for the probabilistic knowledge bases. It introduces problems to calculate these inconsistency measures. Based on the class of inconsistencies, this chapter presents two consistency recovery models: the standard recovery model and the nonstandard recovery model. The construction theorems and inconsistency handling algorithms are also presented in this chapter.

Chapter 4 presents an overview of methods for inconsistency handling and knowledge integration. Selected knowledge integration problems and approaches to conducting knowledge integration are presented. One can find an overview of knowledge integration models, the integration process development and a general principle diagram for knowledge integration systems based on probabilistic knowledge.

In Chapter 5, some distance-based methods for integrating PKBs are proposed. We propose two methods of integrating probabilistic knowledge bases: the method based on the distance and the method based on the probability value. The class of probability integration operators and the properties they should satisfy based on the concept of inconsistency and based on the probability value is introduced. Integration problems, definitions, and theorems to build integrating algorithms are presented in this chapter. Next, the experimental data set, system configuration; analyze and evaluate the experimental results on the reliability of the obtained results, as well as the performance with the basic algorithms proposed in Chapter 6 and the integrating algorithm.

In Chapter 6, some value-based model for integrating PKBs are presented. We introduce two probability value-based integrating operators (MIO, CMIO) and showed that all members satisfy many desired properties that have been considered in the literature. In order to eliminate the redundant probabilistic constraint in the resulting knowledge base, the probabilistic constraint reduction rule is employed so that the reduced knowledge base is still equivalent to the original knowledge base.

Chapter 7 includes the description of the performed experiments. By using a family of DDFs and various norms, we present the evaluation of integration results and investigate how the coefficients in DDFs may affect these results. Several experiments are conducted to evaluate and discuss a number of iterations, CPU times, execution time, etc. by changing the DDFs and norms. These experiments demonstrate the plausibility and feasibility of the proposed algorithms.

In Chapter 7, we present several potential applications of our approach. These proposals show that the proposed methods for knowledge inconsistency resolving and knowledge integration are not only possible, but also effective in using them in concrete practical situations.

Finally, Chapter 8 presents the concluding remarks about the contributions of this book and the discussion about selected open problems.

CHAPTER 2

Probabilistic knowledge-based systems

THE PURPOSE OF THIS CHAPTER is to present the overview of knowledge representation methods, types of knowledge-based systems. In particular, this book introduces a logical organization of a probabilistic knowledge-based system. Components of a probabilistic knowledge-based system also are considered and compared with other systems. One of the most important stages in this architecture is the process for integrating PKBs.

2.1 KNOWLEDGE BASE REPRESENTATION

2.1.1 Knowledge Representation Methods

The knowledge provided by experts is represented in the knowledge base of the system according to a certain format. This is called a knowledge representation of the system. Knowledge representation consists of the coding of the expert knowledge into the system. There exists various types of knowledge-representation schemes suitable for different sorts of knowledge. There are two kinds of knowledge, that is, quantitative knowledge and qualitative knowledge. The first kind is often assigned to different types of heuristics. The second kind is divided into several categories: descriptive knowledge, structural knowledge, procedural knowledge, meta knowledge, heuristic knowledge, and control knowledge. Descriptive knowledge (also known as propositional knowledge, declarative knowledge or constative knowledge) is a basic problem-solving knowledge that can be expressed in a declarative sentence or an indicative proposition. Structural knowledge refers to the knowledge that describes the relationship between concepts and objects. Procedural knowledge can be simply stated as knowing how to perform a specific skill or task, and provides methods for structuring, combining and deducing new knowledge from existing knowledge. Meta knowledge is a fundamental conceptual instrument to define knowledge about other types of knowledge. Heuristic knowledge is seen as some expert knowledge in the field or the subject for experimenting with what the machine has learned. Control knowledge is used to control and combine the different procedural and descriptive knowledge sources. If these types of knowledge do not depend on the space and time factors, they are called the static knowledge. This knowledge source is derived from professional documents, the general science principles. However, knowledge can only be true or false in many cases. This type of knowledge is called uncertain knowledge. An important issue in solving knowledge-based problems is to consider the expert knowledge to extract the heuristic knowledge, the inference rules and the control strategies. Distinguishing types of knowledge is the basis for finding an appropriate knowledge representation method. Knowledge representation is a study of how the beliefs, the intentions, and the judgments of an intelligent agent can be expressed suitably for automated reasoning. Knowledge representation is a knowledge-coding method to represent information from the real world in a form that the knowledge-based systems can understand and then utilize to solve complex real-life problems. Knowledge representation not only stores data in a database but also allows a computer to learn from that knowledge and proceed intelligently like a human being.

The general methods for representing knowledge in knowledge-based systems are detailed in [35] including: Logical representation, semantic network representation, frame representation, production rule representation, Bayesian network representation.

For the convenience of illustrating several concepts related to the following sections, we use the following example:

Example 2.1 *A hospital conducts a survey to evaluate the symptoms of heart disease. Doctors who are surveyed will supply their opinions about symptoms of the heart disease, such as chest pain and shortness of breath. Doctors are asked to make some statements (knowledge) about: the incidence of heart disease, the incidence of shortness of breath, the incidence of chest pain, the incidence of heart disease when this person has symptoms of shortness of breath, the incidence of heart disease when this person has symptoms of chest pain, the incidence of shortness of breath when this person has heart disease, the incidence of chest pain when this person has heart disease. Each doctor will give his or her own knowledge, which is the statement about the incidence of heart disease related to the two symptoms of shortness of breath and chest pain. However, each doctor will give their own knowledge that could also be inconsistent. A requirement of the survey is to provide a general assessment of the* *patient's signs of heart disease from the knowledge of doctors. Although all knowledge of doctors may be consistent, the joint knowledge base may not be consistent. The issues that the survey needs to take into account are:*

1. How to represent the knowledge of each doctor?

2. How to know if the knowledge of a doctor is inconsistent?

3. How to restore the consistency of the knowledge of a doctor?

4. How to integrate the knowledge of the doctors into a joint knowledge that best represents the combined knowledge?

In order to answer the above questions, we first need to find a formal representation of knowledge. The knowledge representation techniques detailed by Harmelen et al. [35] are not suitable for this kind of knowledge. A suitable technique is the probabilistic knowledge representation. The problem of knowledge representation for probabilistic models will be discussed in detail in Section 2.1.2.

2.1.2 Probabilistic Knowledge Base Representation

Let S be a sample space that is the set of all possible outcomes of a statistical experiment. Let $\mathsf{E} = 2^{\mathcal{S}} \setminus \emptyset$ be a finite set of events, where each event is a subset of the sample space S. For example, if the results of a test consist of determining the gender of an infant then $\mathsf{S} = \{g, b\}$, where g means that the child is a girl and b means that the child is a boy. If $E = \{g\}$ then E is an event that baby is a girl. Similarly, if $E = \{b\}$ then E is an event that baby is a boy.

For $F, G \in \mathsf{E}$, the intersection of F and G, denoted by FG, is an event consisting of all joint elements F and G. The negative element of F is defined by $\overline{F} = \mathcal{S} \setminus F$. Let $\mathsf{E} = \{E_1, \ldots, E_n\}$. A complete conjunction Θ of E has form $\Theta = \tilde{E}_1 \tilde{E}_2 \ldots \tilde{E}_n$ with $\tilde{E}_i = \{E_i, \bar{E}_i\}$. Let $\Lambda(\mathsf{E})$ be a set of all complete conjunctions of E so $\Lambda(\mathsf{E}) = \{\Theta_1, \ldots, \Theta_{2^n}\}$. Let $\mathcal{Q} = \mathsf{E} \cup \{\tilde{E}_i \tilde{E}_j | \tilde{E}_i \in \{E_i, \bar{E}_i\}, \tilde{E}_j \in \{E_j, \bar{E}_j\}$, and $E_i \neq E_j \in \mathsf{E}\}$. A complete conjunction $\Theta \in \Lambda(\mathsf{E})$ satisfies $U \in \mathcal{Q}$, denoted by $\Theta \models U$ iff U positively appears in Θ. Let $\mho(U) = \{\Theta \in \Lambda(\mathsf{E}) | \Theta \models U$ and $U \in \mathcal{Q}\}$. Let $\hbar_{\mathsf{E}} = |\Lambda(\mathsf{E})|$ be the number of complete conjunctions E. Let $\mathbb{R}$ be the set of real numbers and $\mathbb{R}^*$ be the set of nonnegative real numbers, i.e. $\mathbb{R}^* = \{x | x \in \mathbb{R}, x \geq 0\}$. Let $\mathbb{R}_{[0,1]}$ be a set of all real numbers in the interval $0 - 1$, i.e. $\mathbb{R}_{[0,1]} = \{x | x \in \mathbb{R}, 0 \leq x \leq 1\}$. Let $\mathbb{R}^n$ be a set of real n-vectors, i.e. $\mathbb{R}^n = \{(x_1, \ldots, x_n) | x_i \in \mathbb{R}\}$. Let $\mathbb{R}^n_{[0,1]}$ be a set of real n-vectors in the interval $0 - 1$, i.e. $\mathbb{R}^n = \{(x_1, \ldots, x_n) | x_i \in \mathbb{R}, 0 \leq x_i \leq 1\}$. Let $\mathbb{R}^{n \times m}$ be a set of real matrices with n rows and m columns.

Definition 2.1 ([36]) *Function* $\mathcal{P} : \Lambda(\mathsf{E}) \rightarrow \mathbb{R}_{[0,1]}$ *is a probability function if*
$$\sum_{i=1}^{\hbar_{\mathsf{E}}} \mathcal{P}(\Theta_i) = 1$$

The following definition represents the set of all probability functions for building problems with integrating PKBs.

Definition 2.2 *Let $\widehat{\mathcal{P}}(\mathsf{E})$ be the set of all probability functions $\mathcal{P}$ over E. The set of all discrete $\hbar_\mathsf{E}$-ary probability functions over E is $\mathbb{P}^{\hbar_\mathsf{E}} = \{(p_1, \ldots, p_m) \in \mathbb{R}_{[0,1]}^{\hbar_\mathsf{E}} | \sum_{i=1}^{\hbar_\mathsf{E}} p_i = 1\}$, where each $p_i \in \vec{p} \in \mathbb{P}^{\hbar_\mathsf{E}}$ corresponds to a $\mathcal{P}(\Theta) \in \widehat{\mathcal{P}}(\mathsf{E})$.*

Let $\vec{\omega} = (\omega_1, \ldots, \omega_{\hbar_\mathsf{E}})^T$ be a column vector, where ω_i corresponds to a probability $\mathcal{P}(\Theta_i)$.

The probability of an event E is the sum of the weights of all sample points in E, denoted by $\mathcal{P}(E)$. Therefore,

$$0 \leq \mathcal{P}(E) \leq 1 \quad \text{and} \quad \mathcal{P}(\mathsf{S}) = 1 \tag{2.1}$$

The conditional probability of F, given the occurrence of some other event G, denoted by $\mathcal{P}(F|G)$, is defined as follows:

$$\mathcal{P}(F|G) = \frac{\mathcal{P}(FG)}{\mathcal{P}(G)} = \rho \text{ provided } \mathcal{P}(G) > 0 \tag{2.2}$$

Two events F and G are independent iff $\mathcal{P}(F|G) = \mathcal{P}(F)$ or $\mathcal{P}(G|F) = \mathcal{P}(G)$ assuming the existences of the conditional probabilities. Otherwise, F and G are dependent.

Theorem 2.1 ([36, 37]) *Let $F, G \in \mathsf{E}$, the probability function $\mathcal{P}$ satisfies the following probability rules:*
(P0) $\mathcal{P}(F) = \sum\limits_{\Theta \in \Lambda(\mathsf{E}):\Theta \models F} \mathcal{P}(\Theta)$
(P1) $\mathcal{P}(FG) = \sum\limits_{\Theta \in \Lambda(\mathsf{E}):\Theta \models FG} \mathcal{P}(\Theta)$, $\mathcal{P}(FG) = \mathcal{P}(F)\mathcal{P}(G|F) = \mathcal{P}(G)\mathcal{P}(F|G)$ provided $\mathcal{P}(F) > 0$, $\mathcal{P}(G) > 0$.
Assume that $\mathcal{P}(F) > 0$ and $\{G_1, ..., G_n\}$ constitute a partition of the sample space S such that $\mathcal{P}(G_k) > 0 \, \forall k = \overline{1, n}$. Then,
(P2) $\mathcal{P}(G_k|F) = \frac{\mathcal{P}(G_k)\mathcal{P}(F|G_k)}{\sum\limits_{i=1}^{n} \mathcal{P}(G_i)\mathcal{P}(F|G_i)} \forall k = \overline{1, n}$
(P3) $\mathcal{P}(F) = \sum\limits_{i=1}^{n} \mathcal{P}(G_i)\mathcal{P}(F|G_i)$
(P4) $\mathcal{P}(F) = 1 - \mathcal{P}(\bar{F})$
If $G_1, G_2, \ldots, G_n$, $n = 1, 2, ..., \infty$ is a sequence of mutually exclusive events, that is, $G_i G_j = \emptyset \, \forall i \neq j$),
(P5) $\mathcal{P}\left(\bigcup\limits_{i=1}^{n} G_i\right) = \sum\limits_{i=1}^{n} \mathcal{P}(G_i)$

Proof 2.1 *Rule **(P0)** can be easily deduced from Definition 2.1. Rule **(P1)** is the result of Definition 2.10, Rule **(P2)** is the result of Theorem 2.14, Rule **(P3)** is the result of Theorem 2.13, Rule **(P4)** is the result of Theorem 2.9, Rule **(P5)** is the result of Corollary 2.2, which has been proposed and proven in [37].* □

Example 2.2 *Let's continue Example 2.1. Let H, T, and D be the events that a patient has heart disease, has difficulty in breathing, and has chest pain, respectively.*

- The set of events is $\mathsf{E} = \{H, T, D\}$ and the set of all complete conjunctions of E is $\Lambda(\mathsf{E}) = \{HTD, HT\bar{D}, H\bar{T}D, H\bar{T}\bar{D}, \bar{H}TD, \bar{H}T\bar{D}, \bar{H}\bar{T}D, \bar{H}\bar{T}\bar{D}\}$. Then, $\hbar_\mathsf{E} = 8$

- The probability of events: Let $\mathcal{P}(H)$, $\mathcal{P}(T)$, and $\mathcal{P}(D)$ be the probabilities that a patient has heart disease, has difficulty in breathing, and has chest pain, respectively.

- Conditional probabilities: Let $\mathcal{P}(T|H)$ and $\mathcal{P}(H|D)$ denote the probabilities that a patient feels short of breath when he or she has heart disease, and a patient has heart disease when he or she feels chest pain, respectively.

A PKB defined by Potyka and Thimm [38, 39, 40] is a set of conditional probabilities considered in a propositional language. The following definitions of a PKB are considered in the probability context, that is, they are considered on the set of events E.

Definition 2.3 *Let $F, G \in \mathsf{E}$ and $\rho \in \mathbb{R}_{[0,1]}$. A probabilistic constraint κ is an expression of the form $c[\rho]$, where $c = (F|G)$.*

Intuitively, $(F|G)[\rho]$ indicates the probability that an event F happens in condition event G hold ρ. We can write $(F)[\rho]$ in the case G is tautological, i.e. $G \equiv \top$. The constraint $(F)[\rho]$ is understood as $\mathcal{P}(F) = \rho$, and it is similar to the probability of an event defined in [37].

Definition 2.4 *A PKB is a finite set of probabilistic constraints, denoted by $\mathcal{K} = \{\kappa_1, \ldots, \kappa_h\}$, where $\kappa_i = c_i[\rho_i], \forall i = \overline{1, h}$.*

Let $\bar{b}_\mathcal{K} = |\mathcal{K}| = h$ be the number of probabilistic constraints in $\mathcal{K}$. Let $\mathbb{K}$ be the set of all PKBs. Let $\mathbb{V}$ be the set of all probability vectors of probabilistic constraints in each $\mathcal{K} \in \mathbb{K}$. Let $\mathsf{SC}(\mathcal{K}) = \{\kappa_1, \ldots, \kappa_{\bar{b}_\mathcal{K}}\}$ be the set of all probabilistic constraints appearing in $\mathcal{K}$. Let $\vec{\lambda} = (\lambda_1, \ldots, \lambda_{\bar{b}_\mathcal{K}})^T$ be a column vector, where an auxiliary variable λ_i corresponds to a probabilistic constraint κ_i. Let $\vec{1} = (1, \ldots, 1)^T$ be a 1-column vector of size $\bar{b}_\mathcal{K}$. Let $\vec{0} = (0, \ldots, 0)^T$ be a 0-column vector of size $\bar{b}_\mathcal{K}$. Let $\vec{\rho}_\mathcal{K} = (\rho_1, \ldots, \rho_{\bar{b}_\mathcal{K}})^T$ be a column vector, where an auxiliary variable ρ_i corresponds to a value of probabilistic constraint κ_i.

Because the knowledge integration process is to search a common knowledge base that best represents a finite set of knowledge bases, it is necessary to define this finite set. In a probabilistic environment, a PKB profile consists of PKBs defined on the same finite set of events.

Definition 2.5 *A PKB profile $\mathcal{R}$ over E is a tuple $\langle \mathcal{B}, \mathsf{E} \rangle$, where:*

1. *$\mathsf{E} = \{E_1, \ldots, E_n\}$ is a finite set including n events.*

2. *$\mathcal{B} = \{\mathcal{K}_1, \ldots, \mathcal{K}_m\}$ is a finite multi-set including m PKBs.*

Let $\hbar_{\mathcal{B}} = |\mathcal{B}| = m$ be the number of PKBs in $\mathcal{B}$. Let $\mathbb{B}$ be a set of finite multi-sets of PKBs.

Example 2.3 *Continuing Example 2.2. The hospital collects knowledge from nine doctors; each doctor is assigned to a code of DT1, DT2, DT3, DT4, DT5, respectively. After the survey, the hospital obtains the knowledge of the doctors shown in Table 2.1.*

Table 2.1   Knowledge of doctors

Probabilistic constraint	DT1	DT2	DT3	DT4	DT5
$\mathcal{P}(H)$	0.7	0.7	0.57	0.7	0.8
$\mathcal{P}(T)$	0.3	0.6	0.69	0.6	0.31
$\mathcal{P}(D)$	0.45	0.5	0.75	0.64	0.5
$\mathcal{P}(H\,|D)$	0.64	0.8	-	0.9	0.6
$\mathcal{P}(T\,|H)$	0.5	0.78	0.67	-	0.42

Let $c_1 = (H)$, $c_2 = (T)$, $c_3 = (D)$, $c_4 = (T\,|H)$, $c_5 = (H\,|D)$. By Definition 2.4, we have:

- $\mathcal{K}_1 = \{c_1[0.7], c_2[0.3], c_3[0.45], c_4[0.5], c_5[0.64]\}$
- $\mathcal{K}_2 = \{c_1[0.7], c_2[0.6], c_3[0.5], c_4[0.78], c_5[0.8]]\}$
- $\mathcal{K}_3 = \{c_1[0.57], c_2[0.69], c_3[0.75], c_4[0.67]\}$
- $\mathcal{K}_4 = \{c_1[0.7], c_2[0.6], c_3[0.64], c_5[0.9]\}$
- $\mathcal{K}_5 = \{c_1[0.8], c_2[0.31], c_3[0.5], c_4[0.42], c_5[0.6]\}$

Therefore, $\bar{b}_{\mathcal{K}_1} = \bar{b}_{\mathcal{K}_2} = \bar{b}_{\mathcal{K}_5} = 5$ and $\bar{b}_{\mathcal{K}_3} = \bar{b}_{\mathcal{K}_4} = 4$. By Definition 2.5, $\mathcal{R} = \langle \mathcal{B}, \mathsf{E} \rangle$ consists of $\mathsf{E} = \{H, T, D\}$, and $\mathcal{B} = \{\mathcal{K}_1, \ldots, \mathcal{K}_5\}$.

2.2 TYPES OF KNOWLEDGE-BASED SYSTEMS

This section summarizes knowledge-based systems that are built on the representation of the knowledge base in the system.

Today, artificial intelligence (AI) has been used in applications to solve specific problems in many fields such as knowledge-based systems, speech recognition, natural language processing, artificial vision, robots, neural networks with specific applications such as personalized shopping, AI-powered assistants, fraud prevention, administrative tasks, automated to aid educators, creating smart content, voice assistants, personalized learning, autonomous vehicles. In AI, each KBS is a computer system with the ability to think and make decision as a human expert. They are designed to deal with a large range of problems from computer science and engineering to social sciences such as economics, politics, and law. There are several well-known knowledge bases being developed and widely applied, namely DBPedia, Google's Knowledge Graph, Neil, Open IE, Probase and Yago. When developing knowledge-based systems, several mathematical models such as Bayesian network models [41, 42] and Markov network models [41, 43] have been selected to represent a knowledge base. Recently, Islam et al. [42] used applications of the fuzzy-Bayesian models to build a KBS for the cost overrun risk assessment of the power plant project. Blondet et al. [44] also use Bayesian networks as an inference engine to build a special KBS. The knowledge base is supported by a specific ontology. The inference engine reasons on the knowledge to assist designers in configuring numerical design of experimental processes efficiently. This system depends on an ontological model to integrate existing expert knowledge and to discover new knowledge from these analyses. The KBS also depends on the inference engine to combine knowledge fast. Enrique et al. [41] proposed typical architecture of a KBS consisting of many of the basic components shown in Figure 2.4.

Two types of problems that KBS could solve are deterministic and stochastic problems. Therefore, KBS can be classified into two main types: deterministic and stochastic KBS [41].

Rule-Based KBSs

In everyday life, many complex situations are governed by deterministic rules: traffic control mechanisms, security systems, banking transactions and so on. The rule-based expert system is an effective tool to solve these problems. Some applications of this type of system are known as train traffic control, the prisoners' problem [41], and CADIAG-2 [45]. These systems will provide users with a list of the concluding facts along with the rules that were used to draw each conclusion.

Probabilistic-Based KBSs

It is not easy to handle uncertainties by using deterministic rule-based KBSs because rules and objects may be deterministically implemented. However, uncertainty is the rule that could exist in KBSs as there may be an imprecise concrete knowledge base or an indeterminate abstract knowledge base. The probability-based KBSs is one of the effective systems to handle the uncertainty of PKB [41].

Probability-based KBSs consist of four main components: PKB, the inference engine, explanatory system, and learning system.

- A PKB includes a set of variables, $\{X_1, \ldots, X_n\}$ and a joint probability distribution over all the variables $p(x_1, \ldots, x_n)$. Therefore, in order to build a PKB, finding the joint probability distribution of the variables is very essential.

- In the inference engine, the conditional probabilities may be evaluated and analyzed by employing methods such as symbolic propagation, approximate propagation and exact propagation [46, 47, 48, 41].

A principle diagram of another KBS is proposed by Chen and Wang [43]. A PKB consisting of entities, facts, rules, events and integrity constraints stored in database tables, $\Gamma = (\mathcal{E}, \mathcal{C}, \mathcal{R}, \Pi, \mathcal{L})$, where $\mathcal{E}$ is a set of entities collating to real-world objects; $\mathcal{C} \subseteq 2^{\mathcal{E}}$ is a set of classes; $\mathcal{R} \subseteq \mathcal{E} \times \mathcal{E}$ is a set of relations; Π is a set of weighted facts (or relationships); and $\mathcal{L}$ is a set of weighted clauses (or rules). Chen and Wang employ the random variables to propose a SQL-based grounding algorithm for finding a relational model for PKBs. This algorithm works effectively by applying Markov's logic network rules in batches. Recently, a new approach reasons on knowledge graphs to obtain new knowledge and conclusions [49]. The new reasoning methods over knowledge graphs in this approach include rule-based reasoning, neural network-based reasoning, and distributed representation-based reasoning. A prototype system, QAS, and a knowledge management tool, QKManager [50], for the quality analysis in model-based design are developed by making full use of the advantages of rule-based KBS. This system can detect model quality defects and offer modification suggestions in the design stage.

In order to focus attention on the advantages of the probabilistic-based KBSs, we will compares the four main periods between the rule-based KBSs and the probabilistic-based KBSs.

(i) **Knowledge Base**: The abstract knowledge base in the rule-based KBSs is the set of rules and the objects, while for the probabilistic-based KBSs, it contains variables, its possible values and their JPD. Otherwise, the concrete knowledge in both systems only consists of the facts.

(ii) **Inference Engine**: The rule-based KBSs employ the inference approaches to obtain conclusions from the facts whereas the probabilistic-based KBSs apply the evaluation of conditional probabilities to acquire the best joint probability distribution. Therefore, the inference engine in the probabilistic-based KBSs is more complex and difficult to perform than that in the rule-based KBSs.

(iii) **Explanation System**: The explanation system of the rule-based KBSs justifies which rule is efficient and partly responsible for the inference process to reach the conclusion while this of the probabilistic-based KBSs depends on the relative values of conditional probabilities, which help to evaluate influences of variables and obtain conclusions.

(iv) **Learning System**: The learning system of the rule-based KBSs builds a set of new objects, a set of new rules by altering or combined the existing objects and rules while that of the probabilistic-based KBSs finds the new joint probability distribution by consolidating or making adjustments to a set of variables or probability values.

2.3 THE KNOWLEDGE-BASED SYSTEM DEVELOPMENT

Although the process of knowledge-based systems development has been promoted to obtain great strides, there exists an extremely time-consuming task. In practice, the cycle for developing the knowledge-based systems could not be divided into smaller components precisely. However, it is necessary to take into account the stages to develop the knowledge-based systems.

Figure 2.1 picks out the five primary stages for the knowledge-based system development. It consists of (i) Identification, (ii) Conceptualization, (iii) Formalization (iv) Implementation, (v) Verification.

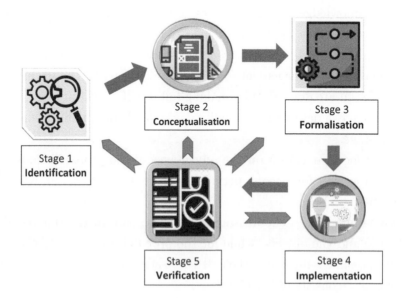

Figure 2.1 The process of a knowledge-based system development.

Stage 1. Identification: Before a knowledge-based system is developed, it is essential to identify the problem, which this system is intended to deal with. The system could provide erroneous inferences if the problem is not defined with as much precision as possible. Therefore, this is most important one in the process of developing a knowledge-based system.

Stage 2. Conceptualization: The problem will be analyzed further to make sure that its specifics as well as generalities are understandable. It is necessary to create the problem diagram for describing the relationships between the objects and processes in the problem domain. This makes the problem able to be divided into sub-problems easier. Therefore, it is easy for the knowledge engineer to find the methods for dealing with the problem.

Stage 3. Formalization: The relationships detected in the conceptualization stage are analyzed to ensure that there exists the solution of the problem. The methods that are appropriate for developing this particular knowledge-based system are considered and chosen by the knowledge engineer.

Stage 4. Implementation: After the formalized concepts as well as their methods have been selected, it can be programmed into the computer. In order to implement knowledge-based systems, it is necessary to make a decision to select a shell, a tool, or a programming language. The knowledge-based system tools or shells offer an advantage for non-computer professionals while the programming languages often provide the most flexibility and the best performance. Then, prototype knowledge-based systems are implemented to test the design and the functions of the systems. If the prototypes do not work, it may be necessary to reformalize the concepts and rechoose the techniques.

Stage 5. Verification: The main purpose of this stage is to verify that the system has been built entirely correctly. It is helpful for the knowledge engineers to identify the limitations on the structure and the implementation of the system. Based on these weaknesses, the appropriate corrections could be considered.

2.4 COMPONENTS OF A PROBABILISTIC KNOWLEDGE-BASED SYSTEM

In the spirit of the architecture of a KBS proposed in [41, 43], our probabilistic KBS consists of four main components as follows:

(1) PKB Building Engine: This component is to build probabilistic constraint-based PKB by using some statistical models in natural language processing or machine learning literature.

(2) Probabilistic Constraint Deducting Engine: This component downsizes the number of probabilistic constraints in a PKB by the probability rules and deduction rules.

(3) Consistency-Restoring Engine: This component employs the method of minimizing the inconsistency measure to allow obtaining a consistent PKB from an inconsistent one.

(4) PKBs Integrating Engine: This component integrates the input PKBs into a joint one by minimizing the divergence distances among them.

The illustration of the proposed architecture is in Figure 2.2.

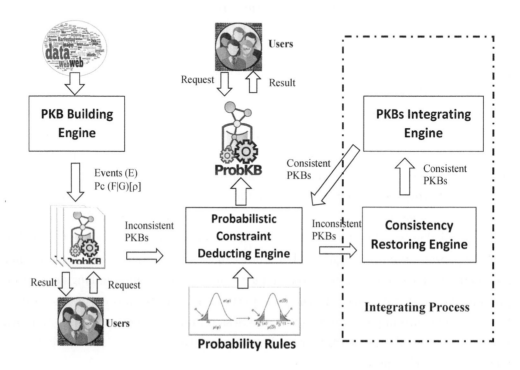

Figure 2.2 Architecture of a probabilistic knowledge-based system [51].

A PKB is built by the PKB Building Engine tool by collecting the knowledge from several knowledge sources, such as experts or knowledge systems. This tool uses several statical methods, such as probabilistic context-free grammars (PCFGs) and log-linear (maximum-entropy) taggers [52]. Next, the PKB is then downsized by the Probabilistic Constraint Deducting Engine tool. This tool works based on the probability rules (Theorem 2.1) and probabilistic constraint deduction rules (Theorem 6.2). Then, if the PKB is inconsistent, it is made consistent by the Consistency Restoring Engine tool. Lastly, some consistent PKBs will be put into the PKBs Merger Engine tool to obtain a joint consistent PKB. The Consistency Restoring Engine and PKBs Merger Engine are major components of the proposed architecture. They make an integration process.

In this book, we define the integration process in our framework with three main stages (as shown in Figure 2.3) as follows:

- **Stage 1**: Calculating inconsistency measures. We will present this stage in detail in Chapter 3.

- **Stage 2**: Solving inconsistences. We will present this stage in detail in Chapter 4.

- **Stage 3**: Integrating PKBs. We will present this stage in detail in Chapter 5 corresponding to the distance-based methods and Chapter 6 corresponding to the value-based method.

We will present the mathematical basis for the proposed integration process in the next subsection. The following is the operation of the integration process: The integration process has input as a tuple of events and PKBs. The inconsistency degree of each PKB is measured by Algorithm 3 in Chapter 3. If the degree is greater than zero, the consistency of the PKB will be restored by using Algorithm 5, Algorithm 6, Algorithm 7, Algorithm 8 in Chapter 4. Algorithm 10 and Algorithm 11 in Chapter 5 are employed to perform the integration process.

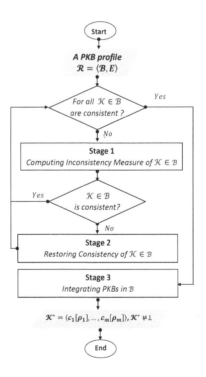

Figure 2.3 The PKBs integration process.

2.5 COMPARING PROBABILISTIC KNOWLEDGE-BASED SYSTEM WITH OTHER SYSTEMS

Figure 2.4 illustrates the typical architecture for a KBS with 10 components as follows.

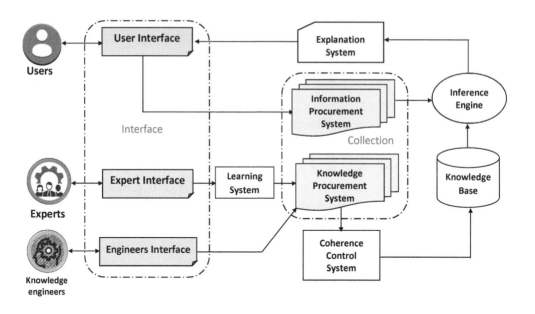

Figure 2.4 Typical architecture of a KBS [41, 51].

The Users Component provides the knowledge base in the specialized domain, this knowledge is translated by the Knowledge Engineers component into a language that can be understood by the Expert System component. The Knowledge Base component consists of two knowledge types, namely, abstract knowledge and concrete knowledge. Abstract knowledge mentions rules or probability distributions; concrete knowledge specifies the information associated with a particular application. The Coherence Control component ensures the consistency of the knowledge base and avoids any inconsistent knowledge added to the knowledge base. Inference Engine component draws conclusions by applying the abstract knowledge to the concrete knowledge. The Inference Engine component uses the Information Procurement System component to obtain knowledge as the output of the inference process. The Interface component provides the working interface between the Expert System component and the users, expert, and knowledge engineers. The Explanation System component gives the explaination for the process followed by the Inference Engine component. Lastly, the Learning System component discovers new knowlege for the system.

Table 2.2 presents the comparison among our KBSs (Figure 2.2 and Figure 2.3) to the existing KBSs (Figure 2.4) in term of tasks, components and methods to implement. Firstly, in our KBS, each knowledge base is a set of probabilistic constraints, while knowledge bases in the existing KBS are represented by JSD. By this changing, we can improve the expressiveness and difficulty of knowledge representation. Concretely, a knowledge base represented by probabilistic constraints is smaller and easy

Table 2.2 Comparison between the proposed KBSs and the existing KBSs [51]

Duty	Existing KBS		Proposed KBS	
	Components	Methods	Components	Methods
Provide the knowledge base	The Human Component	JPD	Several experts, Different sources of statistical data	Probabilistic constraints
Translate knowledge into language of system	Knowledge engineers	Constructing the JPD	PKB Building Engine, Probabilistic Constraint Deducting Engine	Constructing the probabilistic constraints, Probabilistic constraint deducting rules
Control the consistency of the knowledge base	Coherence Control	Compatibility of a set of CPDs	Consistency Restoring Engine	Minimizing inconsistency measure
Draw conclusions	Inference Engine	Conditional probability evaluation methods	PKB Integrating Engine	Minimizing divergence distance
Explain the process	Explanation System	Based on conditional probabilities		Desirable properties, DDFs, Probability rules
Discovery of a new knowledge	Learning System	Change in probabilistic model		Computing the new probability values of constraints based on probability rules

to construct than by JPD. Further, because of the deductive ability of probabilistic constraints, it also downsizes the knowledge base more than JSD. Secondly, for inconsistency detection and resolution function, the existing KBS use the Coherence Control component to check the consistency of a probabilistic model by using the Checking Compatibility algorithm of a set of CPDs. Thus, it can only detect the inconsistency and prevent adding the new inconsistent information to the knowledge base without resolving the existing inconsistency of the knowledge base. In contrast, our KBS implements this function by the Consistency Restoring Engine component. It applies the principle of maximum entropy to minimize inconsistency measures and return a consistent PKB. In this case, the inconsistency measures are determined by solving several optimization problems [39, 40, 53, 54]. Lastly, in our KBS, PKBs Integrating Engine component assumes the roles of components Inference Engine, Explanation System and Learning System in the existing KBS. Concretely, the Inference Engine function is implemeted based on minimizing the divergence distance among joint PKB and the given PKBs. In this case, the lowest divergence distance is the result of resolving integrating problems built on DDFs. The result is given as a new probability distribution function. The explanation function of this component works based on the evaluation of the desirable properties of the integrating operators. The learning system function is the process to take the given probability rules and assess the new probability values of probabilistic constraint. In summary, by these strong improvements, our proposed KBS can overcome several drawbacks of the existing KBS such as inconsistency resolution, input structure limitation, and downsizing of input data. We continue to present and discuss more about the theoretical bases and methods to implement the integration process. in next chapter.

2.6 CONCLUDING REMARKS

In this chapter, we synthesize approaches for building knowledge-based systems such as the Markov network model, Bayesian network model, rule-based expert system, and probabilistic expert system. Based on existing integration methods, we propose an architecture of a probabilistic knowledge-based system and the main stages of the PKBs integration process: consistency restoration stage, PKB integration stage. We also compare the previously integrated system with the probabilistic knowledge-based system.

The theoretical basis and the general model for restoring consistency are detailed in Chapter 4.

The theoretical basis and models for integrating PKBs are detailed in Chapter 5 and Chapter 6.

In the next chapter, we present approaches to compute the inconsistency measure of a PKB.

CHAPTER 3

Inconsistency measures for probabilistic knowledge bases

T HE AIM OF THIS CHAPTER is to introduce the subject of inconsistency of PKBs, the inconsistency representation and how to identify it. We first make a deep survey of the inconsistency measures and present the inconsistency measures for the probabilistic framework. We then investigate a set of desirable properties for inconsistency measures and point out the logical relation among them. Finally, algorithms for computing the inconsistency measures are presented. We also analyze and discuss the complexity of algorithms.

3.1 OVERVIEW OF INCONSISTENCY MEASURES

3.1.1 Distance Functions

This section presents some distance functions that will be used to compute inconsistency measures.

Definition 3.1 ([55]) *Let X be a set. A function $d : X \times X \to \mathbb{R}$ is called a distance function over X if it satisfies the following properties:*

1. $d(x,y) \geq 0$

2. $d(x,y) = 0$ iff $x = y$

3. $d(x,y) = d(y,x)$

4. $d(x,y) + d(y,z) \geq d(x,z)$ where $x,y,z \in X$.

If a distance function only satisfies properties (1)–(3), then it is called a semi-distance function. If it satisfies all properties, it is called a metric. Thus, a distance function measures the distance between two points in a set. Let X be a set and $\sum$ be σ-algebra over X. A function $\mu : \sum \to \mathbb{R}$ is called a measure if it satisfies the following properties: (i) Non-negativity: $\mu(E) \geq 0 \forall E \in \sum$; (ii) Empty set: $\mu(\emptyset) = 0$; (iii) Countable additivity: $\mu(\bigsqcup_{k=1}^{\infty} E_k) = \sum_{k=1}^{\infty} \mu(E_k)$.

For any $\vec{x} \in \mathbb{P}(\mathsf{E})$ and $\vec{y} \in \mathbb{P}(\mathsf{E})$, it is said that $\vec{x}$ dominates $\vec{y}$, denoted by $\vec{x} \gg \vec{y}$, if $x_i = 0$ implies $y_i = 0$. Let $\mathrm{I}(\tilde{\mathrm{x}}) = \{i : x_i \neq 0\}$.

Let $\vec{x} = (x_1, \ldots, x_m) \in \mathbb{R}^m$ and $p \geq 1$ be a real number. By Stephen Boyd and Lieven Vandenberghe [56], p-norm (ℓ_p-norm) of $\vec{x}$ is defined by $\|\vec{x}\|_p = \sqrt[p]{\sum_{i=1}^m |x_i|^p}$. For $p = 1$, 1-norm (ℓ_1-norm, Manhattan norm) is defined by $\|\vec{x}\|_1 = \sum_{i=1}^m |x_i|$. For $p = 2$, 2-norm (ℓ_2-norm, Euclidean norm) is defined by $\|\vec{x}\|_2 = \sqrt{\sum_{i=1}^m |x_i|^2}$. For $p \to \infty$, ∞-norm (ℓ_∞-norm, Maximum norm) is defined by $\|\vec{x}\|_\infty = max\{|x_1|, \ldots |x_m|\}$.

Definition 3.2 ([56]) *Let $\vec{x} = (x_1, \ldots, x_m)$, $\vec{y} = (y_1, \ldots, y_m)$ and $\vec{x}, \vec{y} \in \mathbb{R}^m$. A distance function between two vectors with respect to p-norm ($p \geq 1$) is defined as follows:*

$$d_m^p(\vec{x}, \vec{y}) = \|\vec{x} - \vec{y}\|_p = \sqrt[p]{|x_1 - y_1|^p + \cdots + |x_m - y_m|^p} \tag{3.1}$$

Definition 3.3 ([57]) *Let $\mathcal{K} = \{(F_1 | G_1)[\rho_1], \ldots, \left(F_{\bar{b}_\mathcal{K}} \middle| G_{\bar{b}_\mathcal{K}}\right)[\rho_{\bar{b}_\mathcal{K}}]\}$ be a PKB. Let $\vec{z} = (z_1, \ldots, z_{\bar{b}_\mathcal{K}})^T$, where $z_i = \mathcal{P}(F_i G_i) - \rho_i \mathcal{P}(G_i)$ with $(F_i | G_i)[\rho_i] \in \mathcal{K}$. Distance between a knowledge base $\mathcal{K}$ and a probability function $\mathcal{P}$ with respect to a p-norm is defined as follows:*

$$d^p(\mathcal{K}) = \left\|\left(z_1, \ldots, z_{\bar{b}_\mathcal{K}}\right)\right\|_p = \sqrt[p]{\sum_{i=1}^{\bar{b}_\mathcal{K}} |z_i|^p} \tag{3.2}$$

3.1.2 Development of Inconsistency Measures

Today, one of the most common approaches for dealing with the inconsistencies of knowledge is to use inconsistency measures. An inconsistency measure assigns a non-negative real value to a knowledge base with the intended meaning that the larger the values the larger the inconsistency in knowledge base, with the value equaling zero meaning that knowledge base is consistent. Table 3.1 summarizes techniques for building the inconsistency measures for each type of knowledge base.

Inconsistency measures for logic framework

An overview of measures for classical logics has been proposed in [7, 8, 58, 59, 60, 61]. The idea of the classical measure $\mathcal{I}_d(\mathcal{K})$ [8, 61]. If the knowledge base is consistent, then the inconsistency measure is 1; otherwise it is 0. Inconsistency measures are based on the formulas $\mathcal{I}_c(\mathcal{K})$, $\mathcal{I}_f^d(\mathcal{K})$, $\mathcal{I}_m^p(\mathcal{K})$ [59, 61] interested in the ratio of propositional variables that cause inconsistences. Another approach is that the inconsistency measures $\mathcal{I}_{\mathsf{MI}}(\mathcal{K})$, $\mathcal{I}_{\mathsf{MI}^c}(\mathcal{K})$ [8, 61], $\mathcal{I}_p(\mathcal{K})$ [59, 61] are calculated by counting the number of formulas presenting in the minimal inconsistent subsets. A minimal inconsistent subset of knowledge base could be found by applying the constructive method [62] or the destructive method [63]. In contrast to the approaches proposed in [8, 59, 61], the measures $\mathcal{I}_{\mathsf{mc}}(\mathcal{K})$ [59, 61], $\mathcal{I}_{\mathsf{nc}}(\mathcal{K})$ [58] are defined using maximal inconsistent subsets. A minimal inconsistent subset of knowledge base could be sought directly without finding a minimal inconsistent subset [10]. The measures $\mathcal{I}^h(\mathcal{K})$, $\mathcal{I}^m(\mathcal{K})$, $\mathcal{I}^s(\mathcal{K})$[60, 61] are computed by minimizing the distance function between interpretations in the propositional language. The measure $\mathcal{I}_I^s(\mathcal{K})$ [53] is defined as the sum of the number of minimal inconsistent sets of $\mathcal{K}$. Hunter, Konieczny, and Roussel propose the measure $\mathcal{I}^{si}(\mathcal{K})$ [7, 64] by using the Shapley value to define the degree of inconsistency.

Table 3.1   The inconsistency measures

Techniques	Knowledge Base		
	Logic	Probabilistic-logic	Probability
Classics	$\mathcal{I}_d(\mathcal{K})$ [8, 61]	$\mathcal{I}_0(\mathcal{K})$ [53]	
Formula	$\mathcal{I}_c(\mathcal{K})$, $\mathcal{I}_f^d(\mathcal{K})$ $\mathcal{I}_m^p(\mathcal{K})$ [59, 61]	$\mathcal{I}_\eta(\mathcal{K})$[53]	
The minimal inconsistent subset	$\mathcal{I}_{\mathsf{MI}}(\mathcal{K})$, $\mathcal{I}_{\mathsf{MI}^c}(\mathcal{K})$ [8, 61] $\mathcal{I}_p(\mathcal{K})$ [59, 61]	$\mathcal{I}_{\mathsf{MI}}(\mathcal{K})$, $\mathcal{I}_{\mathsf{MI}}^c(\mathcal{K})$ $\mathcal{I}_p(\mathcal{K})$[53]	
The maximal consistent subset	$\mathcal{I}_{\mathsf{mc}}(\mathcal{K})$ [59, 61] $\mathcal{I}_{\mathsf{nc}}(\mathcal{K})$ [58]		
Distance minimum	$\mathcal{I}^h(\mathcal{K})$, $\mathcal{I}^m(\mathcal{K})$ $\mathcal{I}^s(\mathcal{K})$[60, 61]	$\mathcal{I}_I^s(\mathcal{K})$ $\mathcal{I}_D(\mathcal{K})$ [53],	
Shapley value	$\mathcal{I}^{si}(\mathcal{K})$ [7, 64]		$\mathcal{I}^{si}(\mathcal{K})$ [38, 65]
Optimization problem		$\mathcal{I}^p(\mathcal{K})$ [39, 40] $\mathcal{I}_{\mathcal{IC}}^p(\mathcal{K})$ [40, 66] $\mathcal{I}_x^p(\mathcal{K})$, $\mathcal{I}_\varepsilon^p(\mathcal{K})$ [57]	$\mathcal{I}^*(\mathcal{K})$, $\mathcal{I}^+(\mathcal{K})$ $\mathcal{I}_0^+(\mathcal{K})$ [38, 65] $\mathcal{I}^{ic}$, $\mathcal{I}^{dis}(\mathcal{K})$ $\mathcal{I}^{icoh}(\mathcal{K})$,$\mathcal{I}^{pre}(\mathcal{K})$[34]

Inconsistency measures for probabilistic-logic framework

A class of inconsistency measures for probabilistic-logic framework is detailed in [34, 38, 53, 65, 67]. The measure $\mathcal{I}_0(\mathcal{K})$ [53] differentiates consistent or inconsistent knowledge bases while the measure $\mathcal{I}_\eta(\mathcal{K})$ considers probability functions on probability functions. The measures $\mathcal{I}_{\mathsf{MI}}(\mathcal{K})$, $\mathcal{I}_{\mathsf{MI}^c}(\mathcal{K})$, $\mathcal{I}_p(\mathcal{K})$ [53] could be implemented by enumerating the minimal inconsistent sets, or by combining the numbers of minimal inconsistent sets that contain certain rules. The measure $\mathcal{I}_\eta(\mathcal{K})$ [53] is based on a probability value that it might cause inconsistencies to determine basis inconsistencies. The measure $\mathcal{I}_{\mathsf{MI}}(\mathcal{K})$, $\mathcal{I}_{\mathsf{MI}^c}(\mathcal{K})$, $\mathcal{I}_p(\mathcal{K})$ [53] quantifies inconsistencies by employing the number of minimal inconsistent subsets of a knowledge base. Thimm [53] also suggested several inconsistency measures based on the distance minimization. The inconsistency measure $\mathcal{I}_D(\mathcal{K})$ is defined as a distance that minimizes the distance between a given knowledge base and a consistent knowledge base having the same structure as the input knowledge base. The measure $\mathcal{I}_I^s(\mathcal{K})$ [53] constructed as the sum of the measures $\mathcal{I}_D(\mathcal{M})$, where $\mathcal{M}$ is the minimal inconsistent sets.

Potyka [39, 40] and Bona [57] analyze the inconsistencies based on the distance measure idea p-norm to propose the minimal violation inconsistency measures with respect to 1-norm $\mathcal{I}^1(\mathcal{K})$, p-norm $\mathcal{I}^p(\mathcal{K})$, and ∞-norm $\mathcal{I}^\infty(\mathcal{K})$. These measures correspond to convex optimization problems. Bona [57] suggests inconsistency measures $\mathcal{I}_x^p(\mathcal{K})$ and $\mathcal{I}_\varepsilon^p(\mathcal{K})$ for probabilistic bases based on distance minimization. Bona considers $\mathcal{I}_x^p(\mathcal{K})$ as distance between probability bounds and $\mathcal{I}_\varepsilon^p(\mathcal{K})$ as distance to being satisfied. These measures are computationally superior to those proposed by Thimm [53] and Picado-Muiño [67] because they are based on the principle of maximum entropy. Potyka and Thimm [40, 66] proposes a family of minimal violation measures $\mathcal{I}_{\mathcal{IC}}^p(\mathcal{K})$ with integrity constraint $\mathcal{IC}$. Potyka also points out that all members of this family satisfy a set of desirable properties if $\mathcal{IC} = \emptyset$ thì $\mathcal{I}_{\mathcal{IC}}^p(\mathcal{K}) = \mathcal{I}^p(\mathcal{K})$.

Inconsistency measures for probabilistic framework

The Shapley values also are utilized by Thimm [38, 65] to determine $\mathcal{I}^{si}(\mathcal{K})$, but the formulas in $\mathcal{K}$ are probabilistic constraints. Thimm [38] continues the idea of Picado-Muiño [67] and proposes the inconsistency measure MinDev $\mathcal{I}^*(\mathcal{K})$. Thimm [38, 65] supposes that inconsistencies are caused by the probability value of certain probabilistic constraint in a PKB. Based on this idea, Thimm proposes the measure $\mathcal{I}^+(\mathcal{K})$. Based on the measure $\mathcal{I}^+(\mathcal{K})$, Thimm defines another measure $\mathcal{I}_0^+(\mathcal{K})$. This class of measures is the solution of the optimization problem. Daniel [34] proposes an inconsistent knowledge representation technique using a candidacy function, which assigns a real number between 0 and 1 to each probability function. Daniel introduces an inconsistency measure $\mathcal{I}^{ic}(\mathcal{K})$, dissimilarity measure $\mathcal{I}^{dis}(\mathcal{K})$, incoherence measure $\mathcal{I}^{icoh}(\mathcal{K})$ and precision measure $\mathcal{I}^{pre}(\mathcal{K})$ to detect inconsistencies in knowledge content and be understood as the sum of contradictions within the content of $\mathcal{K}$.

3.2 REPRESENTING THE INCONSISTENCY OF THE PROBABILISTIC KNOWLEDGE BASE

3.2.1 Basic Notions

In this section, we present some notions needed for use in the next sections.

The task of merging knowledge is very important when it is desired to combine several knowledge-based systems into one or to make them interoperable. One of the necessary conditions for successful cooperation is the unity of knowledge of these systems. However, without the ability to merge knowledge, the cooperation between systems is also impossible. This is a difficult problem that needs attention. Therefore, determining the consistency of each knowledge base in the integration process is very essential. The inconsistency measure is one of the basic approaches to represent the inconsistency of a knowledge base. Inconsistency measure is one of the effective tools to analyze and resolve inconsistencies.

First, we consider the consistency of PKBs through the following definition:

Definition 3.4 ([40]) *A probability function* $\mathcal{P} \in \widehat{\mathcal{P}}(\mathsf{E})$ *fulfills a constraint* $(F \,|\, G)\,[\rho]$, *denoted by* $\mathcal{P} \models (F \,|\, G)\,[\rho]$, *iff* $\mathcal{P}(FG) = \rho \mathcal{P}(G)$.

So $\mathcal{P} \models (F)[\rho]$, iff $\mathcal{P}(F) = \rho$.

Definition 3.5 ([40]) *A probability function* $\mathcal{P}$ *fulfills* $\mathcal{K}$, *denoted by* $\mathcal{P} \models \mathcal{K}$, *iff* $P \models \kappa \; \forall \kappa \in \mathcal{K}$. *Let* $\mho(\mathcal{K}) = \{\mathcal{P} \in \widehat{\mathcal{P}}(\mathsf{E}) | \mathcal{P} \models \mathcal{K}\}$ *be a set of all probability functions fulfilling* $\mathcal{K}$.

Let $\mathcal{P}_{\mathcal{K}} : \Lambda(\mathsf{E}) \to \mathbb{R}_{[0,1]}$ be a probability function of $\mathcal{K}$ over $\Lambda(\mathsf{E})$ such that $\sum_{\Theta \in \Lambda(\mathsf{E})} \mathcal{P}_{\mathcal{K}}(\Theta) = 1$. Let $\tilde{\mathcal{P}} : \mathbb{P}^{\hbar_\mathsf{E}} \to \mathbb{R}_{[0,1]}$ be a probability function in $\mathbb{P}^{\hbar_\mathsf{E}}$ such that $\tilde{\mathcal{P}}(\mathcal{P}) > 0$ only with a finite set $\mathcal{P} \in \mathbb{P}^{\hbar_\mathsf{E}}$. Let $\tilde{\Lambda}(\mathsf{E})$ be a set of all probability functions $\tilde{\mathcal{P}}$, defined by $\tilde{\Lambda}(\mathsf{E}) = \{\tilde{\mathcal{P}}(\mathcal{P}) | \tilde{\mathcal{P}}(\mathcal{P}) > 0, \mathcal{P} \in \mathbb{P}(\mathsf{E})\}$.

The probability of constraint κ that fulfills κ is defined as follows:

$$\tilde{\mathcal{P}}(\kappa) = \sum_{\mathcal{P} \in \widehat{\mathcal{P}}(\mathsf{E}), \mathcal{P} \models \kappa} \tilde{\mathcal{P}}(\mathcal{P}) \tag{3.3}$$

Definition 3.6 ([40]) *A PKB* $\mathcal{K}$ *is consistent, denoted by* $\mathcal{K} \not\models \bot$, *iff* $\mho(\mathcal{K}) \neq \emptyset$. *Otherwise,* $\mathcal{K}$ *is inconsistent, denoted by* $\mathcal{K} \models \bot$.

If PKBs are inconsistent, it is necessary to evaluate the inconsistency degree of these PKBs. It is necessary to represent and compute the inconsistency measure of a PKB. Definition 3.7 is suggested to represent the inconsistency measure of a PKB.

Definition 3.7 *Let $\mathcal{R} = \langle \mathcal{B}, \mathbf{E} \rangle$ be a PKB profile. Inconsistency measure $\mathcal{I}$ of $\mathcal{K} \in \mathcal{B}$ over $\mathbf{E}$ is $\mathcal{I} : \mathbb{K} \to \mathbb{R}^*$ such that $\mathcal{I}(\mathcal{K}) = 0$ iff $\mho(\mathcal{K}) \neq \emptyset$, $\mathcal{K} \in \mathbb{K}$.*

A PKB is consistent if there is at least an instance that satisfies all of its formulas, that is, $\mathcal{K}$ is consistent if there is at least a probability function $\mathcal{P}_{\mathcal{K}}$ such that $\mathcal{P} \models \mathcal{K}$. Otherwise, $\mathcal{K}$ is inconsistent if there is no $\mathcal{P}$ such that $\mathcal{P} \models \mathcal{K}$.

Example 3.1 *Consider a PKB in example 2.3. As $\mathcal{P} \models \langle (H)[0.7], (T|H)[0.5]\rangle$, it is easy to see that $\mho(\mathcal{K}_1) = \emptyset$ so $\mathcal{P}(T) \geq 0.7 \times 0.5 = 0.35$, which cannot concurrently be satisfied with $\mathcal{P}(T) = 0.3$. Hence, $\mathcal{K}_1 \models \bot$. Similarly, $\mathcal{K}_5 \models \bot$, that is, $\mathcal{I}(\mathcal{K}_1) \neq 0$ and $\mathcal{I}(\mathcal{K}_5) \neq 0$. However, we have $\mho(\mathcal{K}_2) = \mho(\mathcal{K}_3) = \mho(\mathcal{K}_4) \neq \emptyset$, $\mathcal{K}_2 \not\models \bot$, $\mathcal{K}_3 \not\models \bot$, $\mathcal{K}_4 \not\models \bot$, that is, $\mathcal{I}(\mathcal{K}_2) = \mathcal{I}(\mathcal{K}_3) = \mathcal{I}(\mathcal{K}_4) = 0$.*

If $\mathcal{K}$ is inconsistent, then the minimal inconsistent subsets of $\mathcal{K}$ could be defined as follows:

Definition 3.8 ([57]) *A set of probabilistic constraints $\mathcal{M}^i \subseteq \mathcal{K}$ is minimal inconsistent subset if $\mathcal{M}^i \models \bot$ and $\forall \mathcal{M}_s^i \subset \mathcal{M}^i$ then $\mathcal{M}_s^i \not\models \bot$. Then, the set of the minimal inconsistent subsets of $\mathcal{K} \in \mathbb{K}$ is defined as follows:*

$$\mathsf{SMI}(\mathcal{K}) = \{\mathcal{M}^i \subseteq \mathcal{K} | \mathcal{M}^i \models \bot, \mathcal{M}_s^i \subset \mathcal{M}^i : \mathcal{M}_s^i \not\models \bot\} \tag{3.4}$$

Definition 3.9 ([57]) *A set of probabilistic constraints $\mathcal{M}^c \subseteq \mathcal{K}$ is a maximal consistent subset if $\mathcal{M}^c \not\models \bot$ and $\forall \mathcal{M}^c \subset \mathcal{M}_s^c \subseteq \mathcal{K}$ then $\mathcal{M}_s^c \models \bot$. Then, be the set of the maximal consistent subsets of $\mathcal{K} \in \mathbb{K}$ is defined as follows:*

$$\mathsf{SMC}(\mathcal{K}) = \{\mathcal{M}^c \subseteq \mathcal{K} | \mathcal{M}^c \not\models \bot, \forall \mathcal{M}^c \subset \mathcal{M}_s^c \subseteq \mathcal{K} : \mathcal{M}_s^c \models \bot\} \tag{3.5}$$

Definition 3.10 ([57]) *The set of self-contradictory constraints of $\mathcal{K} \in \mathbb{K}$ is defined as follows:*

$$\mathsf{SCC}(\mathcal{K}) = \{\kappa \in \mathcal{K} | \kappa \models \bot\} \tag{3.6}$$

Definition 3.11 ([57]) *A probabilistic constraint $\kappa \in \mathcal{K}$ is a free constraint iff $\kappa \notin \mathcal{M}^i \, \forall \mathcal{M}^i \in \mathsf{SMI}(\mathcal{K})$. Then, the set of all free constraints of $\mathcal{K}$ is defined as follows:*

$$\mathsf{Fc}(\mathcal{K}) = \{\kappa \in \mathcal{K} | \kappa \notin \mathcal{M}^i, \forall \mathcal{M}^i \in \mathsf{SMI}(\mathcal{K})\} \tag{3.7}$$

It is easy to see that $\mathsf{Fc}(\mathcal{K}) = \bigcup_{\mathcal{M}^i \in \mathsf{SMI}(\mathcal{K})} \mathcal{M}^i$

Let $\mathsf{App}(\mathfrak{S})$ be a set of events appearing in $\mathfrak{S}$, where $\mathfrak{S}$ is a PKB $\mathcal{K}$, a constraint κ, or a complete conjunction Θ.

Definition 3.12 ([57]) *A probabilistic constraint $\kappa \in \mathcal{K}$ is a safe constraint in $\mathcal{K}$ iff* $App(\kappa) \cap App(\mathcal{K} \backslash \{\kappa\}) = \emptyset$. *Then, the set of all safe constraints of $\mathcal{K}$ is defined as follows:*

$$Sc(\mathcal{K}) = \{\kappa \in \mathcal{K} | App(\kappa) \cap App(\mathcal{K} \backslash \{\kappa\}) = \emptyset\} \tag{3.8}$$

Definition 3.13 *A PKB profile $\mathcal{R} = \langle \mathcal{B}, E \rangle$ is a consistent PKB profile iff $\forall \mathcal{K}_i \in \mathcal{B} : \mathcal{K}_i \not\models \perp$.*

3.2.2 Characteristic Model

To build a model for solving the consistency-restoring problem and the problems with integrating PKBs, it is necessary to build a characteristic model of PKB. The characteristic model of each PKB is determined by the characteristic matrix, the non-negative coefficient matrix, the non-positive coefficient matrix, the double diagonal matrix, and the characteristic function. The following definition is suggested to represent the relationship between events and complete conjunctions.

Definition 3.14 *An indicate function $\delta : \mathcal{Q} \times \Lambda(E) \to \mathbb{R}_{[0,1]}$ is defined as follows:*

$$\delta(U, \Theta) = \begin{cases} 1 & if \ \Theta \models U. \\ 0 & otherwise. \end{cases} \tag{3.9}$$

The following definition is employed to represent the matrix forms of a PKB.

Definition 3.15 *Let $\mathcal{R} = \langle \mathcal{B}, E \rangle$ be a PKB profile. A characteristic matrix of $\mathcal{K} \in \mathcal{B}$ over E is defined as follows:*

$$A_{\mathcal{K}}^{E} = (a_{ij}) \in \mathbb{R}^{\bar{b}_{\mathcal{K}} \times \hbar_E} \tag{3.10}$$

A non-negative coefficient matrix of $\mathcal{K} \in \mathcal{B}$ over E is defined as follows:

$$C_{\mathcal{K}}^{E,+} = (c_{ij}^{+}) \in \mathbb{R}^{\bar{b}_{\mathcal{K}} \times \hbar_E} \tag{3.11}$$

A non-positive coefficient matrix of $\mathcal{K} \in \mathcal{B}$ over E is defined as follows:

$$C_{\mathcal{K}}^{E,-} = (c_{ij}^{-}) \in \mathbb{R}^{\bar{b}_{\mathcal{K}} \times \hbar_E} \tag{3.12}$$

A double diagonal matrix of $\mathcal{K} \in \mathcal{B}$ over E is defined as follows:

$$\bar{A}_{\mathcal{K}} = (\bar{a}_{ij}) \in \mathbb{R}^{\bar{b}_{\mathcal{K}} \times 2\bar{b}_{\mathcal{K}}} \tag{3.13}$$

where $a_{ij} = \delta\left(F_iG_i, \Theta_j\right)(1 - \rho_i) - \delta\left(\overline{F_iG_i}, \Theta_j\right)\rho_i$, $c_{ij}^+ = \delta\left(F_iG_i, \Theta_j\right)$, $c_{ij}^- = \delta\left(G_i, \Theta_j\right)$, and

$$
\bar{a}_{ij} = \begin{cases} 1 & \text{if } i = j \text{ and } i = \overline{1, \bar{b}_{\mathcal{K}}} \\ -1 & \text{if } j - i = \bar{b}_{\mathcal{K}} \text{ and } j = \overline{\bar{b}_{\mathcal{K}} + 1, 2\bar{b}_{\mathcal{K}}} \\ 0 & \text{otherwise} \end{cases}
$$

Let $\vec{a}_j$ be the i-th row vector of matrix $A_{\mathcal{K}}^{\mathsf{E}}$. Let $B_{\mathcal{K}}^{\mathsf{E}} = (\vec{a}_j)^T = (A_{\mathcal{K}}^{\mathsf{E}})^T = (b_{ij}) \in \mathbb{R}^{\hbar_{\mathsf{E}} \times \bar{b}_{\mathcal{K}}}$.

Example 3.2 *Consider $\mathcal{K}_1$ in Example 2.3. By Definition 3.14, 3.15 etc., we have*
$a_{11} = \delta\left(H, HTD\right)0.3 - \delta\left(\bar{H}, HTD\right)0.7 = 0.3$; *similarly,* $a_{12} = a_{13} = a_{14} = 0.3$,
$a_{21} = a_{22} = a_{25} = a_{26} = 0.7$, $a_{31} = a_{33} = a_{35} = a_{37} = 0.55$, $a_{41} = a_{43} = 0.5$,
$a_{51} = a_{53} = 0.36$; $a_{15} = \delta\left(H, \bar{H}TD\right)0.3 - \delta\left(\bar{H}, \bar{H}TD\right)0.7 = -0.7$; *similarly,* $a_{16} = $
$a_{17} = a_{18} = -0.7$, $a_{23} = a_{24} = a_{27} = a_{28} = -0.3$, $a_{32} = a_{34} = a_{36} = a_{38} = $
-0.45, $a_{42} = a_{44} = -0.5$, $a_{55} = a_{57} = -0.64$; *we have* $a_{45} = \delta\left(HT, \bar{H}TD\right)0.5 - $
$\delta\left(\bar{H}T, \bar{H}TD\right)0.5 = 0$; *similarly,* $a_{46} = a_{47} = a_{48} = a_{52} = a_{54} = a_{56} = a_{58} = 0$.
Therefore, we have:

$$
A_{\mathcal{K}_1}^{\mathsf{E}} = \begin{pmatrix}
0.3 & 0.3 & 0.3 & 0.3 & -0.7 & -0.7 & -0.7 & -0.7 \\
0.7 & 0.7 & -0.3 & -0.3 & 0.7 & 0.7 & -0.3 & -0.3 \\
0.55 & -0.45 & 0.55 & -0.45 & 0.55 & -0.45 & 0.55 & -0.45 \\
0.5 & -0.5 & 0.5 & -0.5 & 0 & 0 & 0 & 0 \\
0.36 & 0 & 0.36 & 0 & -0.64 & 0 & -0.64 & 0
\end{pmatrix}
$$

By Definition 3.14, 3.15 etc., we also have

$$
C_{\mathcal{K}_1}^{\mathsf{E},+} = \begin{pmatrix}
1 & 1 & 1 & 1 & 0 & 0 & 0 & 0 \\
1 & 1 & 0 & 0 & 1 & 1 & 0 & 0 \\
1 & 0 & 1 & 0 & 1 & 0 & 1 & 0 \\
1 & 1 & 0 & 0 & 0 & 0 & 0 & 0 \\
1 & 0 & 1 & 0 & 0 & 0 & 0 & 0
\end{pmatrix}, C_{\mathcal{K}_1}^{\mathsf{E},-} = \begin{pmatrix}
-1 & -1 & -1 & -1 & -1 & -1 & -1 & -1 \\
-1 & -1 & -1 & -1 & -1 & -1 & -1 & -1 \\
-1 & -1 & -1 & -1 & -1 & -1 & -1 & -1 \\
-1 & -1 & -1 & -1 & 0 & 0 & 0 & 0 \\
-1 & 0 & -1 & 0 & -1 & 0 & -1 & 0
\end{pmatrix}
$$

$$
\bar{A}_{\mathcal{K}_1} = \begin{pmatrix}
1 & 0 & 0 & 0 & 0 & -1 & 0 & 0 & 0 & 0 \\
0 & 1 & 0 & 0 & 0 & 0 & -1 & 0 & 0 & 0 \\
0 & 0 & 1 & 0 & 0 & 0 & 0 & -1 & 0 & 0 \\
0 & 0 & 0 & 1 & 0 & 0 & 0 & 0 & -1 & 0 \\
0 & 0 & 0 & 0 & 1 & 0 & 0 & 0 & 0 & -1
\end{pmatrix}
$$

Definition 3.16 is employed to represent the characteristic function of a PKB.

Definition 3.16 *Let* $\mathcal{K} = \{(F_1|G_1)[\rho_1], \ldots, (F_h|G_h)[\rho_h]\}$. *Function* $\partial_{\mathcal{K}} : \mathbb{V} \to \mathbb{K}$ *is called the characteristic function of* $\mathcal{K}$ *if there exists* $\vec{\vartheta} = (\vartheta_1, \ldots, \vartheta_h) \in \mathbb{V}$ *such that* $\partial_{\mathcal{K}}(\vec{\vartheta}) = \{(F_1|G_1)[\vartheta_1], \ldots, (F_h|G_h)[\vartheta_h]\}$.

Intuitively, Definition 3.16 states that the characteristic function maps a value vector interval in $[0,1]$ to a PKB. The input vector values are substitutes for the corresponding value of probabilistic constraints in the PKB.

Definition 3.17 ([40]) *Let* $\mathcal{R} = \langle \mathcal{B}, \mathsf{E} \rangle$ *be a PKB profile and* $\mathcal{K}_1, \mathcal{K}_2 \in \mathcal{B}$

- *Two probabilistic constraints* $\kappa_1, \kappa_2 \in \mathcal{K}$ *are equivalent, denoted by* $\kappa_1 \equiv \kappa_2$, *iff* $\mho(\{\kappa_1\}) = \mho(\{\kappa_2\})$.

- *Two PKBs* $\mathcal{K}_1, \mathcal{K}_2$ *are extensionally equivalent, denoted by* $\mathcal{K}_1 \overset{\Delta}{=} \mathcal{K}_2$, *iff* $\mho(\mathcal{K}_1) = \mho(\mathcal{K}_2)$.

- *Two PKBs* $\mathcal{K}_1, \mathcal{K}_2$ *are semi-extensionally equivalent, denoted by* $\mathcal{K}_1 \overset{\wedge}{=} \mathcal{K}_2$, *if* $\exists$ *a bijection* $\alpha : \mathcal{K}_1 \to \mathcal{K}_2$ *such that* $\kappa \equiv \alpha(\kappa) \ \forall \kappa \in \mathcal{K}_1$.

- *Two PKBs* $\mathcal{K}_1, \mathcal{K}_2$ *are qualitatively equivalent, denoted by* $\mathcal{K}_1 \cong \mathcal{K}_2$, *iff* $\bar{b}_{\mathcal{K}_1} = \bar{b}_{\mathcal{K}_2}$ *and* $\exists \vec{\beta} \in \mathbb{V} : \mathcal{K}_1 = \partial_{\mathcal{K}_2}(\vec{\beta})$.

- *Two PKBs* $\mathcal{K}_1, \mathcal{K}_2$ *are in same structure if* $\forall \mathcal{K}_i, \mathcal{K}_j \ (i \neq j) \in \mathcal{B} : \bar{b}_{\mathcal{K}_i} = \bar{b}_{\mathcal{K}_j} = h$ *and* $\forall (F_k|G_k) [\rho_k] \in \mathcal{K}_i$, $(F_{k'}|G_{k'}) [\rho_{k'}] \in \mathcal{K}_j$, $k = k' : (F_k|G_k) \simeq (F_{k'}|G_{k'})$.

3.2.3   Desired Properties of Inconsistency Measures

In knowledge-based systems, the concept of consistency of a knowledge base is often understood as a situation in which a knowledge base does not normally contain contradictions. The degree of inconsistency of a knowledge base is usually determined through inconsistency measures. Therefore, it is important to define the properties that the inconsistency measures should satisfy. This ensures the reliability and the accuracy of inconsistency measures. In this section, several properties of $\mathcal{I}$ are used to characterize inconsistency measures. Some of the following properties are derived and inherited from the logic model and the probabilistic-logic model [7, 57]. They are improved to fit the probabilistic model.

Definition 3.18 ([7, 57]) *(The desirable property of inconsistency measures). Let* $\mathcal{K}$, $\mathcal{K}_1$ *and* $\mathcal{K}_2$ *be PKBs,* κ *be a probabilistic constraint. Inconsistency measure* $\mathcal{I} : \mathbb{K} \to \mathbb{R}^*$ *should satisfy the following desirable properties:*

(CON)-Consistency. $\mathcal{K} \not\models \bot$ *iff* $\mathcal{I}(\mathcal{K}) = 0$.

The property **CON** *represents the minimal requirement for an inconsistency measure.*

(MON)-Monotonicity. $\mathcal{I}(\mathcal{K}) \leq \mathcal{I}(\mathcal{K} \cup \kappa)$.

> The property **MON** shows that if a probabilistic constraint is added to the PKB, the value of inconsistency measure will increase.

(SUA)-Super-Additivity. If $\mathcal{K}_1 \cap \mathcal{K}_2 = \emptyset$ then $\mathcal{I}(\mathcal{K}_1 \cup \mathcal{K}_2) \geq \mathcal{I}(\mathcal{K}_1) + \mathcal{I}(\mathcal{K}_2)$.

> The property **SUA** requires that the inconsistency measure of the union of two PKBs will be greater than the sum of inconsistency measure of each PKB. It is easy to see that **SUA** is stronger than **MON** because **SUA** can be viewed as a special case of **MON**.

(NOR)-Normalization. $\mathcal{I}(\mathcal{K}) \in [0,1]$.

> The property **NOR** shows that the inconsistency measure is always in the unit interval.

(MIS)- MIS-Separability. If $SMI(\mathcal{K}_1 \cup \mathcal{K}_2) = SMI(\mathcal{K}_1) \cup SMI(\mathcal{K}_2)$ and $SMI(\mathcal{K}_1) \cap SMI(\mathcal{K}_2) = \emptyset$ then $\mathcal{I}(\mathcal{K}_1 \cup \mathcal{K}_2) = \mathcal{I}(\mathcal{K}_1) + \mathcal{I}(\mathcal{K}_2)$.

> The property **MIS** states that if the minimal consistent subsets of one knowledge base is separate from those of the other, the inconsistency value of uninion of two knowledge bases is the sum of the inconsistency values of those.

(FCI)-Free-constraint independence. If $\kappa \in Fc(\mathcal{K})$ then $\mathcal{I}(\mathcal{K}) = \mathcal{I}(\mathcal{K} \backslash \{\kappa\})$. The property **FCI** states that if a free constraint is removed from $\mathcal{K}$, its inconsistency measure is not changed.

(SCI)-Safe-constraint independence. If $\kappa \in Sc(\mathcal{K})$ then $\mathcal{I}(\mathcal{K}) = \mathcal{I}(\mathcal{K} \backslash \{\kappa\})$.

> The property **SCI** states that if a safe constraint is removed from $\mathcal{K}$, its inconsistency measure is not changed.

(IRS)-Irrelevance of syntax. If $\mathcal{K}_1 \overset{\triangle}{=} \mathcal{K}_2$ then $\mathcal{I}(\mathcal{K}_1) = \mathcal{I}(\mathcal{K}_2)$.

> The property **IRS** states that if $\mathcal{K}_1$ is semi-extensionally equivalent to $\mathcal{K}_2$, the inconsistency measure of $\mathcal{K}_1$ is equal to that of $\mathcal{K}_2$.

(PEN)-Penalty. If $\kappa \notin Fc(\mathcal{K})$ and $\kappa \in \mathcal{K}$ then $\mathcal{I}(\mathcal{K}) > \mathcal{I}(\mathcal{K} \backslash \{\kappa\})$.

> The property **PEN** states that the addition of the inconsistent information makes the inconsistency measure increased.

Definition 3.19 ([7, 57]) *If the inconsistency measure of a probabilistic knowledge base satisfies properties* **CON, NOR, MON, FCI**, *it is called a basic inconsistency measure.*

3.3 INCONSISTENCY MEASURES FOR PROBABILISTIC KNOWLEDGE BASES

3.3.1 The Basic Inconsistency Measures

Definition 3.20 ([8]) *The drastic inconsistency measure* $\mathcal{I}_{dr} : \mathbb{K} \to \mathbb{R}^*$ *of* $\mathcal{K}$ *is defined as follows:*

$$\mathcal{I}_{dr}(\mathcal{K}) = \begin{cases} 1, & \text{if } \mathcal{K} \models \bot. \\ 0, & \text{otherwise.} \end{cases} \tag{3.14}$$

Theorem 3.1 ([8]) $\mathcal{I}_{dr}$ *fulfills* **CON, MON, NOR, FCI, IRS** *and* **SCI**.

Proof 3.1 *(CON).* *By Definition 3.6, we have* $\mathcal{K} \not\models \bot$ *iff* $\mho(\mathcal{K}) \neq \emptyset$. *Moreover, by Definition 3.7,* $\mathcal{I}(\mathcal{K}) = 0$ *iff* $\mho(\mathcal{K}) \neq \emptyset$.

(MON). *If* $\mathcal{K} \not\models \bot$ *then* $\mathcal{I}_{dr}(\mathcal{K}) = 0 \leq \mathcal{I}_{dr}(\mathcal{K} \cup \kappa)$.
If $\mathcal{K} \models \bot$ *then* $\mathcal{I}_{dr}(\mathcal{K}) = 1 = \mathcal{I}_{dr}(\mathcal{K} \cup \kappa)$.

(NOR). $\forall \mathcal{K} : \mathcal{I}_{dr}(\mathcal{K}) = 0$ *or* $\mathcal{I}_{dr}(\mathcal{K}) = 1$. *Hence,* $\mathcal{I}_{dr}(\mathcal{K}) \in [0,1]$.

(FCI). *Assume that* $\mathcal{K} \not\models \bot$. *As* $\kappa \in \mathsf{Fc}(\mathcal{K})$, *by Definition 3.11* $\kappa \notin \mathcal{M}^i$ $\forall \mathcal{M}^i \in \mathsf{SMI}(\mathcal{K})$. *Hence,* $\mathcal{K} \backslash \{\kappa\} \not\models \bot$. *Therefore,* $\mathcal{I}_{dr}(\mathcal{K}) = \mathcal{I}_{dr}(\mathcal{K} \backslash \{\kappa\}) = 0$.

Assume that $\mathcal{K} \models \bot$. *We also* $\mathcal{K} \backslash \{\kappa\} \models \bot$. *Therefore,* $\mathcal{I}_{dr}(\mathcal{K}) = \mathcal{I}_{dr}(\mathcal{K} \backslash \{\kappa\}) = 1$

(SCI). *As* $\kappa \in \mathsf{Sc}(\mathcal{K})$, *by Proposition 3 [53]* $\kappa \in \mathsf{Fc}(\mathcal{K})$. *Hence,* $\mathcal{I}_{dr}(\mathcal{K}) = \mathcal{I}_{dr}(\mathcal{K} \backslash \{\kappa\})$.

(IRS). *If* $\mathcal{K}_1 \overset{\Delta}{=} \mathcal{K}_2$, $\mathcal{K}_1 \overset{\Delta}{=} \mathcal{K}_2$. *By Definition 3.17,* $\mho(\mathcal{K}_1) = \mho(\mathcal{K}_2)$. *Therefore,* $\mathcal{I}_{dr}(\mathcal{K}_1) = \mathcal{I}_{dr}(\mathcal{K}_1)$. $\qquad \square$

It is easy to see that $\mathcal{I}_{dr}$ violates **SUA, MIS, PEN**, and it is a basic inconsistency measure.

Example 3.3 *Consider the knowledge bases* $\mathcal{K}_1$ *and* $\mathcal{K}_5$ *in example 2.3.*

It follows that $\mathcal{I}_{dr}(\mathcal{K}_1) = \mathcal{I}_{dr}(\mathcal{K}_5) = 1$ *and* $\mathcal{I}_{dr}(\mathcal{K}_2) = \mathcal{I}_{dr}(\mathcal{K}_3) = \mathcal{I}_{dr}(\mathcal{K}_4) = 0$.

It follows that $\mathcal{K}_1 \cap \mathcal{K}_5 = \emptyset$ *and* $\mathcal{I}_{dr}(\mathcal{K}_1 \cup \mathcal{K}_5) = 1 < \mathcal{I}_{dr}(\mathcal{K}_1) + \mathcal{I}_{dr}(\mathcal{K}_5) = 2$ *thus* $\mathcal{I}_{dr}$ *of* $\mathcal{K}_1$ *and* $\mathcal{K}_5$ *fails to satisfy* **SUA**.

As $\mathsf{SMI}(\mathcal{K}_1) = \{\{(H)[0.7], (T)[0.3], (T|H)[0.5]\}\}$ *and* $\mathsf{SMI}(\mathcal{K}_5) = \{\{(H)[0.8], (T)[0.31], (T|H)[0.42]\}\}$, $\mathcal{I}_{dr}$ *of* $\mathcal{K}_1$ *and* $\mathcal{K}_5$ *fails to satisfy* **MIS**.

Moreover, $\mathcal{I}_{dr}$ *of* $\mathcal{K}_1$ *and* $\mathcal{K}_5$ *fails to satisfy* **PEN** *as* $\mathsf{Fc}(\mathcal{K}_1) = \emptyset$ *and* $\mathsf{Fc}(\mathcal{K}_5) = \emptyset$.

Definition 3.21 ([8, 53]) *The MI-inconsistency measure* $\mathcal{I}_{mi} : \mathbb{K} \to \mathbb{R}^*$ *of* $\mathcal{K}$ *is defined as follows:*

$$\mathcal{I}_{mi}(\mathcal{K}) = |\mathsf{SMI}(\mathcal{K})| \tag{3.15}$$

Theorem 3.2 ([8, 53]) $\mathcal{I}_{mi}$ *fulfills CON, MON, SUA, MIS, FCI, SCI, IRS, and PEN.*

Proof 3.2 *(CON). If $\mathcal{K} \not\models \perp$, we have SMI$(\mathcal{K}) = \emptyset$. Therefore, $\mathcal{I}_{mi}(\mathcal{K}) = 0$.*

(MON). We have $|$SMI$(\mathcal{K})| \leq |$SMI$(\mathcal{K} \cup \{\kappa\})|$. Hence, $\mathcal{I}_{mi}(\mathcal{K}) \leq \mathcal{I}_{mi}(\mathcal{K} \cup \{\kappa\})$.

(SUA). Since $\mathcal{K}_1 \cap \mathcal{K}_2 = \emptyset$ it follows that SMI$(\mathcal{K}_1) \subseteq$ SMI$(\mathcal{K}_1 \cup \mathcal{K}_2)$ and SMI$(\mathcal{K}_2) \subseteq$ SMI$(\mathcal{K}_1 \cup \mathcal{K}_2)$. Due to $\mathcal{K}_1 \cap \mathcal{K}_2 = \emptyset$ it holds that SMI$(\mathcal{K}_1) \cap$ SMI$(\mathcal{K}_2) = \emptyset$. Hence, SMI$(\mathcal{K}_1) \cup$ SMI$(\mathcal{K}_2) \subseteq$ SMI$(\mathcal{K}_1 \cup \mathcal{K}_2)$. Therefore, $\mathcal{I}_{mi}(\mathcal{K}_1 \cup \mathcal{K}_2) = |SMI(\mathcal{K}_1 \cup \mathcal{K}_2)| \geq |SMI(\mathcal{K}_1) \cup$ SMI$(\mathcal{K}_2)| = |$SMI$(\mathcal{K}_1)| + |$SMI$(\mathcal{K}_2)| = \mathcal{I}_{mi}(\mathcal{K}_1) + \mathcal{I}_{mi}(\mathcal{K}_2)$.

(MIS). $\mathcal{I}_{mi}(\mathcal{K}_1 \cup \mathcal{K}_2) = |SMI(\mathcal{K}_1 \cup \mathcal{K}_2)| = |SMI(\mathcal{K}_1)| + |SMI(\mathcal{K}_2)| = \mathcal{I}_{mi}(\mathcal{K}_1) + \mathcal{I}_{mi}(\mathcal{K}_2)$.

(IRS). As $\mathcal{K}_1 \stackrel{\triangle}{=} \mathcal{K}_2$, due to Definition 3.17 it holds that $\exists$ a bijection $\alpha : \mathcal{K}_1 \rightarrow \mathcal{K}_2$ such that $\kappa \equiv \alpha(\kappa) \; \forall \kappa \in \mathcal{K}_1$. Let $\mathcal{K}^* \subseteq \mathcal{K}_1$ and $\alpha(\mathcal{K}^*) = \{\alpha(\kappa) | \kappa \in \mathcal{K}^*\}$. Hence, by Definition 3.17, $\mho(\{\kappa\}) = \mho(\{\alpha(\kappa)\}) \; \forall \kappa \in \mathcal{K}_1$. It is easy to see that $\mathcal{M}^i \in$ SMI$(\mathcal{K}_1)$ iff $\alpha(\mathcal{M}^i) \in$ SMI$(\mathcal{K}_2)$, that is, $|$SMI$(\mathcal{K}_1)| = |$SMI$(\mathcal{K}_2)|$. Therefore, $\mathcal{I}_{mi}(\mathcal{K}_1) = \mathcal{I}_{mi}(\mathcal{K}_2)$.

(FCI). Since $\kappa \in$ Fc$(\mathcal{K})$, by Definition 3.11 $\kappa \notin \mathcal{M}^i \; \forall \mathcal{M}^i \in$ SMI$(\mathcal{K})$. Hence, $|$SMI$(\mathcal{K})| = |$SMI$(\mathcal{K}\backslash\{\kappa\})|$. Therefore, $\mathcal{I}_{mi}(\mathcal{K}) = \mathcal{I}_{mi}(\mathcal{K}\backslash\{\kappa\})$.

(SCI). As $\kappa \in$ Sc$(\mathcal{K})$, by Proposition 3 [53] $\kappa \in$ Fc$(\mathcal{K})$. Hence, $\mathcal{I}_{mi}(\mathcal{K}) = \mathcal{I}_{mi}(\mathcal{K}\backslash\{\kappa\})$.

(PEN). Since $\kappa \notin$ Fc$(\mathcal{K})$, by Definition 3.11 $\kappa \in \mathcal{M}^i \; \forall \mathcal{M}^i \in$ SMI$(\mathcal{K})$. Hence, SMI$(\mathcal{K}\backslash\{\kappa\}) \subseteq$ SMI$(\mathcal{K})$. We then obtain $\mathcal{I}_{mi}(\mathcal{K}\backslash\{\kappa\}) = |SMI(\mathcal{K}\backslash\{\kappa\})| \leq |SMI(\mathcal{K})| = \mathcal{I}_{mi}(\mathcal{K})$. □

Example 3.4 *Consider $\mathcal{K}_1$ and $\mathcal{K}_5$ in example 2.3. By formula 3.15, $\mathcal{I}_{mi}(\mathcal{K}_1) = \mathcal{I}_{mi}(\mathcal{K}_2) = 1$. As $\mathcal{I}_{mi}(\mathcal{K}_1 \cup \mathcal{K}_2) = 2 \notin [0,1]$, $\mathcal{I}_{mi}$ of $\mathcal{K}_1$ and $\mathcal{K}_2$ fails to satisfy NOR.*

Definition 3.22 ([8, 53]) *An inconsistency measure $\mathcal{I}_{smi}^c : \mathbb{K} \rightarrow \mathbb{R}^*$ of $\mathcal{K}$ is defined as follows:*

$$\mathcal{I}_{smi}^c(\mathcal{K}) = \sum_{\mathcal{M}^i \in SMI(\mathcal{K})} \frac{1}{|\mathcal{M}^i|} \tag{3.16}$$

Note that if SMI$(\mathcal{K}) = \emptyset$, $\mathcal{I}_{smi}^c(\mathcal{K}) = 0$.

Theorem 3.3 ([8, 53]) $\mathcal{I}_{smi}^c$ *fulfills CON, MON, SUA, MIS, FCI, SCI, and PEN.*

Proof 3.3 *(CON). If $\mathcal{K} \not\models \perp$, we have SMI$(\mathcal{K}) = \emptyset$. Therefore, $\mathcal{I}_{smi}^c(\mathcal{K}) = 0$.*

(MON). It holds that SMI$(\mathcal{K}) \subseteq$ SMI$(\mathcal{K} \cup \{\kappa\})$. Hence, $\mathcal{I}_{smi}^c(\mathcal{K} \cup \{\kappa\}) =$

$$\sum_{\mathcal{M}^i \in SMI(\mathcal{K} \cup \{\kappa\})} \frac{1}{|\mathcal{M}^i|} \geq \sum_{\mathcal{M}^i \in SMI(\mathcal{K})} \frac{1}{|\mathcal{M}^i|} + \sum_{\mathcal{M}^i \in SMI(\{\kappa\})} \frac{1}{|\mathcal{M}^i|} \geq \sum_{\mathcal{M}^i \in SMI(\mathcal{K})} \frac{1}{|\mathcal{M}^i|} = \mathcal{I}_{smi}^c(\mathcal{K})$$

(SUA). Since $\mathcal{K}_1 \cap \mathcal{K}_2 = \emptyset$ it follows that SMI$(\mathcal{K}_1) \subseteq$ SMI$(\mathcal{K}_1 \cup \mathcal{K}_2)$ and SMI$(\mathcal{K}_2) \subseteq$

$\mathsf{SMI}(\mathcal{K}_1 \cup \mathcal{K}_2)$. Due to $\mathcal{K}_1 \cap \mathcal{K}_2 = \emptyset$ it holds that $\mathsf{SMI}(\mathcal{K}_1) \cap \mathsf{SMI}(\mathcal{K}_2) = \emptyset$. Hence, $\mathsf{SMI}(\mathcal{K}_1) \cup \mathsf{SMI}(\mathcal{K}_2) \subseteq \mathsf{SMI}(\mathcal{K}_1 \cup \mathcal{K}_2)$. Therefore, $\mathcal{I}_{smi}^c(\mathcal{K}_1 \cup \mathcal{K}_2) = \sum_{\mathcal{M}^i \in \mathsf{SMI}(\mathcal{K}_1 \cup \mathcal{K}_2)} \frac{1}{|\mathcal{M}^i|}$
$\geq \sum_{\mathcal{M}^i \in \mathsf{SMI}(\mathcal{K}_1)} \frac{1}{|\mathcal{M}^i|} + \sum_{\mathcal{M}^i \in \mathsf{SMI}(\mathcal{K}_2)} \frac{1}{|\mathcal{M}^i|} = \mathcal{I}_{smi}^c(\mathcal{K}_1) + \mathcal{I}_{smi}^c(\mathcal{K}_2)$.

(MIS). $\mathcal{I}_{smi}^c(\mathcal{K}_1 \cup \mathcal{K}_2) = \sum_{\mathcal{M}^i \in \mathsf{SMI}(\mathcal{K}_1 \cup \mathcal{K}_2)} \frac{1}{|\mathcal{M}^i|} = \sum_{\mathcal{M}^i \in \mathsf{SMI}(\mathcal{K}_1)} \frac{1}{|\mathcal{M}^i|} + \sum_{\mathcal{M}^i \in \mathsf{SMI}(\mathcal{K}_2)} \frac{1}{|\mathcal{M}^i|}$
$= \mathcal{I}_{smi}^c(\mathcal{K}_1) + \mathcal{I}_{smi}^c(\mathcal{K}_2)$.

(IRS). It is easy to see that $\mathcal{M}^i \in \mathsf{SMI}(\mathcal{K}_1)$ iff $\alpha(\mathcal{M}^i) \in \mathsf{SMI}(\mathcal{K}_2)$, that is, $|\mathsf{SMI}(\mathcal{K}_1)| = |\mathsf{SMI}(\mathcal{K}_2)|$. Hence, $|\mathcal{M}^i| = |\alpha(\mathcal{M}^i)|$. Therefore, $\mathcal{I}_{smi}^c(\mathcal{K}_2) = \sum_{\mathcal{M}^i \in \mathsf{SMI}(\mathcal{K}_2)} \frac{1}{|\mathcal{M}^i|} = \sum_{\mathcal{M}^i \in \mathsf{SMI}(\mathcal{K}_1)} \frac{1}{|\alpha(\mathcal{M}^i)|} = \sum_{\mathcal{M}^i \in \mathsf{SMI}(\mathcal{K}_1)} \frac{1}{|\mathcal{M}^i|} = \mathcal{I}_{smi}^c(\mathcal{K}_1)$.

(FCI). Since $\kappa \in \mathsf{Fc}(\mathcal{K})$, by Definition 3.11 $\kappa \notin \mathcal{M}^i$ $\forall \mathcal{M}^i \in \mathsf{SMI}(\mathcal{K})$. Hence, $\mathsf{SMI}(\mathcal{K}) = \mathsf{SMI}(\mathcal{K} \backslash \{\kappa\})$. Therefore, $\mathcal{I}_{smi}^c(\mathcal{K}) = \sum_{\mathcal{M}^i \in \mathsf{SMI}(\mathcal{K})} \frac{1}{|\mathcal{M}^i|} = \sum_{\mathcal{M}^i \in \mathsf{SMI}(\mathcal{K} \backslash \{\kappa\})} \frac{1}{|\mathcal{M}^i|} = \mathcal{I}_{smi}^c(\mathcal{K} \backslash \{\kappa\})$.

(SCI). As $\kappa \in \mathsf{Sc}(\mathcal{K})$, by Proposition 3 [53] $\kappa \in \mathsf{Fc}(\mathcal{K})$. Hence, $\mathcal{I}_{smi}^c(\mathcal{K}) = \mathcal{I}_{smi}^c(\mathcal{K} \backslash \{\kappa\})$.

(PEN). Since $\kappa \notin \mathsf{Fc}(\mathcal{K})$, by Definition 3.11 $\kappa \in \mathcal{M}^i$ $\forall \mathcal{M}^i \in \mathsf{SMI}(\mathcal{K})$. Hence, $\mathsf{SMI}(\mathcal{K} \backslash \{\kappa\}) \subseteq \mathsf{SMI}(\mathcal{K})$. We then obtain $\mathcal{I}_{smi}^c(\mathcal{K} \backslash \{\kappa\}) = \sum_{\mathcal{M}^i \in \mathsf{SMI}(\mathcal{K} \backslash \{\kappa\})} \frac{1}{|\mathcal{M}^i|} \leq \sum_{\mathcal{M}^i \in \mathsf{SMI}(\mathcal{K})} \frac{1}{|\mathcal{M}^i|} = \mathcal{I}_{smi}^c(\mathcal{K})$. □

Example 3.5 Consider $\mathcal{K}_1$ and $\mathcal{K}_5$ in example 2.3. By example 3.3, $\mathsf{SMI}(\mathcal{K}_1) = \{\{(H)[0.7], (T)[0.3], (T|H)[0.5]\}\}$, $|\mathcal{M}^i| = 3$. Therefore, by formula 3.3.1, $\mathcal{I}_{smi}^c(\mathcal{K}_1) \approx 0.33$. Similarly, $\mathcal{I}_{smi}^c(\mathcal{K}_5) \approx 0.33$.

Definition 3.23 ([61]) An inconsistency measure $\mathcal{I}_\ell : \mathbb{K} \to \mathbb{R}^*$ of $\mathcal{K}$ is defined as follows:

$$\mathcal{I}_\ell(\mathcal{K}) = 1 - \max\{\ell | \exists \tilde{\mathcal{P}} \in \tilde{\Gamma}(E) : \forall \kappa \in \mathcal{K} : \tilde{\mathcal{P}}(\kappa) \geq \ell\} \qquad (3.17)$$

Theorem 3.4 ([61]) $\mathcal{I}_\ell$ fulfills CON, MON, NOR, FCI, SCI, and IRS.

Proof 3.4 (CON). Let $\mathcal{P}$ be a probability function with $\mathcal{P} \models \mathcal{K}$. Let $\tilde{\mathcal{P}}(\mathcal{P}_1) = 1$ and $\tilde{\mathcal{P}}(\mathcal{P}_2) = 0$ $\forall$ $\mathcal{P}_2 \in \tilde{\Gamma}(E)$. Since $\forall$ $\kappa \in \mathcal{K} : \tilde{\mathcal{P}}(\kappa) = 1$ it follows that $\mathcal{I}_\ell(\mathcal{K}) = 1 - 1 = 0$.

(MON). Let $\tilde{\mathcal{P}} \in \tilde{\Gamma}(E)$ and $\ell^* \in [0,1]$ such that $\mathcal{I}_\ell(\mathcal{K} \cup \{\kappa\}) = 1 - \ell^*$ and $\ell^* = \max\{\ell | \exists \tilde{\mathcal{P}} \in \tilde{\Gamma}(E) : \forall \kappa \in \mathcal{K} : \tilde{\mathcal{P}}(\kappa) \geq \ell\}$. It follows that $\forall \kappa \in \mathcal{K} : \tilde{\mathcal{P}}(\kappa) \geq \ell^*$. Therefore, $\mathcal{I}_\ell(\mathcal{K}) = 1 - \max\{\ell | \exists \tilde{\mathcal{P}} \in \tilde{\Gamma}(E) : \forall \kappa \in \mathcal{K} : \tilde{\mathcal{P}}(\kappa) \geq \ell\} \leq 1 - \ell^* = \mathcal{I}_\ell(\mathcal{K} \cup \{\kappa\})$.

(NOR). Since $\tilde{\mathcal{P}} \in [0,1]$ it follows that $\max\{\ell | \exists \tilde{\mathcal{P}} \in \tilde{\Gamma}(E) : \forall \kappa \in \mathcal{K} : \tilde{\mathcal{P}}(\kappa) \geq \ell\} \in [0,1]$. Therefore, $\mathcal{I}_\ell(\mathcal{K}) \in [0,1]$.

(FCI). The proof of the property FCI is analogous to the proof of Proposition 10 in [53].

*(SCI). As $\kappa \in$ **Sc**$(\mathcal{K})$, by Proposition 3 [53] $\kappa \in$ **Fc**$(\mathcal{K})$. Hence, $\mathcal{I}_\ell(\mathcal{K}) = \mathcal{I}_\ell(\mathcal{K} \backslash \{\kappa\})$.*

(IRS). As $\mathcal{K}_1 \stackrel{\wedge}{=} \mathcal{K}_2$, due to Definition 3.17 it holds that $\exists$ a bijection $\alpha : \mathcal{K}_1 \to \mathcal{K}_2$ such that $\kappa \equiv \alpha(\kappa) \ \forall \kappa \in \mathcal{K}_1$. Since $\mho(\kappa) = \mho(\alpha(\kappa)) \ \forall \kappa \in \mathcal{K}_1$ it follows that $\forall \tilde{\mathcal{P}} \in \tilde{\Gamma}(E)$: $\tilde{\mathcal{P}}(\kappa) = \tilde{\mathcal{P}}(\alpha(\kappa))$. Hence, $\mathcal{I}_\ell(\mathcal{K}_1) = \mathcal{I}_\ell(\mathcal{K}_2)$. □

It is easy to see that $\mathcal{I}_\ell$ is a basic inconsistency measure.

Example 3.6 *Consider $\mathcal{K}_1$ and $\mathcal{K}_5$ in example 2.3. For $\mathcal{K}_1$, let*

$\mathcal{K}_{11} = \{(H)[0.7], ((T \,|H)[0.5]\}$,
$\mathcal{K}_{12} = \{(H)[0.7], (T)[0.3], (H|D)[0.64]\}$,
$\mathcal{K}_{13} = \{(H)[0.7], (T \,|H)[0.5], (H|D)[0.64]\}$,
$\mathcal{K}_{14} = \{(H)[0.7], (T)[0.3]\}$,
$\mathcal{K}_{15} = \{(H)[0.7], (D)[0.64], (T|H)[0.5], (H|D)[0.64]\}$,
$\mathcal{K}_{16} = \{(H)[0.7], (D)[0.45], (T)[0.3], (H|D)[0.64]\}$,
$\mathcal{K}_{17} = \{(H)[0.7], (T)[0.3], (D)[0.45]$, $\mathcal{K}_{18} = \{(H)[0.7], (D)[0.64], (T \,|H)[0.5]\}$,

By Definition 3.4, 3.5 etc., $\mathcal{P}_{\mathcal{K}_{11}} \models \mathcal{K}_{11}, \ldots, \mathcal{P}_{\mathcal{K}_{18}} \models \mathcal{K}_{18}$.

Therefore, $\tilde{\mathcal{P}}(\mathcal{P}_{\mathcal{K}_{11}}) = \cdots = \tilde{\mathcal{P}}(\mathcal{P}_{\mathcal{K}_{18}}) = 0.125$.

By formula 3.3, $\tilde{\mathcal{P}}((H)[0.7]) = 8 \times 0.125 = 1$, $\tilde{\mathcal{P}}((T)[0.3]) = \tilde{\mathcal{P}}((T)[0.3]) = 5 \times 0.125 = 0.625$, $\tilde{\mathcal{P}}((T \,|H)[0.5]) = \tilde{\mathcal{P}}((H|D)[0.64]) = 4 \times 0.125 = 0.5$.

Therefore, by formula 3.17, $\mathcal{I}_\ell(\mathcal{K}_1) = 1 - 0.5 = 0.5$. Similarly, $\mathcal{I}_\ell(\mathcal{K}_5) = 0.5$.

*However, $\mathcal{I}_\ell(\mathcal{K}_1 \cup \mathcal{K}_5) = 0.5 < \mathcal{I}_\ell(\mathcal{K}_1) + \mathcal{I}_\ell(\mathcal{K}_5) = 1$. Therefore, $\mathcal{I}_\ell$ of $\mathcal{K}_1$ fails to satisfy **SUA** and **MIS**. Similarly, in example 3.3, $\mathcal{I}_\ell$ of $\mathcal{K}_1$ and $\mathcal{K}_5$ fails to satisfy **PEN**.*

Definition 3.24 ([59]) *An inconsistency measure $\mathcal{I}_\chi : \mathbb{K} \to \mathbb{R}^*$ of $\mathcal{K}$ is defined as follows:*

$$\mathcal{I}_\chi(\mathcal{K}) = |\mathbf{SMC}(\mathcal{K})| + |\mathbf{SCC}(\mathcal{K})| - 1 \tag{3.18}$$

Theorem 3.5 ([59]) *$\mathcal{I}_\chi$ fulfills **CON**, **MON**, **FCI**, **SCI**, and **IRS**.*

Proof 3.5 *(CON). If $\mathcal{K} \not\models \bot$, we have $\mathbf{SMC} = \{\mathcal{K}\}$ and $\mathbf{SCC} = 0$. Therefore, $\mathcal{I}_\chi = 1 + 0 - 1 = 0$.*

(MON). Since $|\mathbf{SCC}(\mathcal{K} \cup \{\kappa\})| \geq |\mathbf{SCC}(\mathcal{K})|$ it follows that $\mathcal{I}_\chi(\mathcal{K} \cup \{\kappa\}) = |\mathbf{SMC}(\mathcal{K} \cup \{\kappa\})| + |\mathbf{SCC}(\mathcal{K} \cup \{\kappa\})| - 1 \geq |\mathbf{SMC}(\mathcal{K})| + |\mathbf{SCC}(\mathcal{K})| - 1 = \mathcal{I}_\chi(\mathcal{K})$

(FCI). By Definition 3.9, we have $\mathbf{SMC}(\mathcal{K}) = \{\mathcal{M}^c \subseteq \mathcal{K} | \mathcal{M}^c \not\models \bot, \forall \mathcal{M}^c \subset \mathcal{M}_s^c \subseteq \mathcal{K} : \mathcal{M}_s^c \models \bot\}$ and $\mathbf{SMC}(\mathcal{K} \backslash \{\kappa\}) = \{\mathcal{M}^{c^*} \subseteq \mathcal{K} \backslash \{\kappa\} | \mathcal{M}^{c^*} \not\models \bot, \forall \mathcal{M}^{c^*} \subset \mathcal{M}_s^{c^*} \subseteq \mathcal{K} \backslash \{\kappa\} : \mathcal{M}_s^{c^*} \models \bot\}$. Since $\kappa \in \mathbf{Fc}(\mathcal{K})$, by Definition 3.11 $\kappa \notin \mathcal{M}^i \ \forall \mathcal{M}^i \in \mathbf{SMI}(\mathcal{K})$.

It follows that $\kappa \notin \mathcal{M}_s^c$ and $\kappa \notin \mathcal{M}_s^{c^*}$. Hence, $|SMC(\mathcal{K})| = |SMC(\mathcal{K} \backslash \{\kappa\})|$. Similarly, $|SCC(\mathcal{K})| = |SCC(\mathcal{K} \backslash \{\kappa\})|$. Therefore, $\mathcal{I}_\chi(\mathcal{K}) = \mathcal{I}_\chi(\mathcal{K} \backslash \{\kappa\})$

(SCI). As $\kappa \in Sc(\mathcal{K})$, by Proposition 3 [53] $\kappa \in Fc(\mathcal{K})$. Hence, $\mathcal{I}_\ell(\mathcal{K}) = \mathcal{I}_\ell(\mathcal{K} \backslash \{\kappa\})$.

(IRS). By Definition 3.9, 3.17 etc., $\forall \kappa \in \mathcal{M}_1^c \subseteq \mathcal{K}_1$: $\exists$ a bijection $\alpha : \mathcal{M}_1^c \to \mathcal{M}_2^c$ such that $\kappa \equiv \alpha(\kappa)$ with $\mathcal{M}_2^c \subseteq \mathcal{K}_2$. Hence, $\mathcal{M}_1^c \subseteq \mathcal{M}_2^c \ \forall \mathcal{M}_1^c \in SMC(\mathcal{K}_1)$ and $\mathcal{M}_2^c \in SMC(\mathcal{K}_2)$. Similarly, $\mathcal{M}_2^c \subseteq \mathcal{M}_1^c$. It follows that $\mathcal{M}_1^c = \mathcal{M}_2^c \ \forall \mathcal{M}_1^c \in SMC(\mathcal{K}_1)$ and $\mathcal{M}_2^c \in SMC(\mathcal{K}_2)$. Therefore, $SMC(\mathcal{K}_1) = SMC(\mathcal{K}_2)$. Similarly, $SCC(\mathcal{K}_1) = SCC(\mathcal{K}_2)$. We have $\mathcal{I}_\chi(\mathcal{K}_1) = |SMC(\mathcal{K}_1)| + |SCC(\mathcal{K}_1)| - 1 = |SMC(\mathcal{K}_2)| + |SCC(\mathcal{K}_2)| - 1 = \mathcal{I}_\chi(\mathcal{K}_2)$. $\square$

Example 3.7 *Consider $\mathcal{K}_1$ and $\mathcal{K}_5$ in example 2.3. By formula 3.5 and from example 3.6, $SMC = \{\mathcal{K}_{11}, \ldots, \mathcal{K}_{18}\}$. By formula 3.6, $SCC = \emptyset$. Therefore, by formula 3.18, $\mathcal{I}_\chi(\mathcal{K}_1) = 8 + 0 - 1 = 7$. Similarly, $\mathcal{I}_\chi(\mathcal{K}_5) = 7$. As $\mathcal{I}_\chi(\mathcal{K}_1) = \mathcal{I}_\chi(\mathcal{K}_5) = 7 \notin [0, 1]$, $\mathcal{I}_\chi$ of $\mathcal{K}_1$ and $\mathcal{K}_5$ fails to satisfy NOR.*

Definition 3.25 *([58]) An inconsistency measure $\mathcal{I}_\mu : \mathbb{K} \to \mathbb{R}^*$ of $\mathcal{K}$ is defined as follows:*

$$\mathcal{I}_\mu(\mathcal{K}) = \bar{b}_\mathcal{K} - max\{\xi | \forall \mathcal{M}^c \subseteq \mathcal{K} : |\mathcal{M}^c| = \xi : \mathcal{M}^c \not\models \perp\} \qquad (3.19)$$

Theorem 3.6 *([58]) $\mathcal{I}_\mu$ fulfills CON, MON, SUA, MIS, FCI, SCI and PEN.*

Proof 3.6 *The proof of the above theorem is analogous to the proof for measures $\mathcal{I}_\ell$ (Theorem 3.4).* $\square$

Example 3.8 *Consider $\mathcal{K}_1$ and $\mathcal{K}_5$ in example 2.3. Observe that $\bar{b}_{\mathcal{K}_1} = 5$ and $max\{\bar{b}_{\mathcal{K}_{11}}, \ldots, \bar{b}_{\mathcal{K}_{18}}\} = 4$ implies $\mathcal{I}_\mu(\mathcal{K}_1) = 5 - 4 = 1$ and similarly $\mathcal{I}_\mu(\mathcal{K}_5) = 1$.*

Definition 3.26 *([7]) An inconsistency measure $\mathcal{I}_{dm}^p : \mathbb{K} \to \mathbb{R}^*$ of $\mathcal{K}$ is defined as follows:*

$$\mathcal{I}_{dm}^p(\mathcal{K}) = \inf\{d_{\bar{b}_\mathcal{K}}^p(\vec{x}, \vec{y}) | \mathcal{K} = \partial_\mathcal{K}(\vec{x}), \partial_\mathcal{K}(\vec{y}) \not\models \perp\} \qquad (3.20)$$

Theorem 3.7 *([7]) $\mathcal{I}_{dm}^p$ fulfills CON, MON, FCI, IRS, and SCI. If $p = 1$ then $\mathcal{I}_{dm}^p$ fulfills SUA.*

Proof 3.7 *(CON). If $\mathcal{K} \not\models \perp$, we have $d_{\bar{b}_\mathcal{K}}^p(\vec{x}, \vec{x}) = 0$ and $d_{\bar{b}_\mathcal{K}}^p(\vec{x}, \vec{y}) \geq 0$. Therefore, $\mathcal{I}_{dm}^p(\mathcal{K}) = 0$.*

(MON). Let $\mathcal{K} = \{\kappa_1, \ldots, \kappa_h\}$, where $\kappa_i = c_i[\rho_i], \forall i = \overline{1, h}$. We have $\vec{y} = (\rho_1, \ldots, \rho_h)$. Assume that $\mathcal{K} \not\models \perp$, there exists $\rho_{h+1} \in [0, 1]$ such that $\mathcal{K}^* \not\models \perp$ where $\mathcal{K}^* = \partial_\mathcal{K}(\vec{y}^*)$ with $\vec{y}^* = \{\rho_1, \ldots, \rho_{h+1}\}$. Let $\mathcal{K} = \partial_\mathcal{K}(\vec{x}) \ \forall \vec{x} = \{\rho_1, \ldots, \rho_h\}$. Hence,

there exists $\rho_{h+1} \in [0,1]$ such that $\mathcal{K}^* = \partial_\mathcal{K}(\vec{x}^*)$ with $\vec{x}^* = \{\rho_1, \ldots, \rho_{h+1}\}$. It follows that $d^p_{\bar{b}_\mathcal{K}}(\vec{x}, \vec{y}) \leq d^p_{\bar{b}_{\mathcal{K}^*}}(\vec{x}^*, \vec{y}^*)$ by Definition 3.2. Therefore, $\mathcal{I}^p_{dm}(\mathcal{K}) \leq \mathcal{I}^p_{dm}(\mathcal{K} \cup \kappa)$.

(IRS). Let $\mathcal{K}_1 = \partial_\mathcal{K}(\vec{x}_1)$, $\mathcal{K}_2 = \partial_\mathcal{K}(\vec{x}_2)$, and $\bar{b}_{\mathcal{K}_1} = \bar{b}_{\mathcal{K}_2} = h$ with $\forall \vec{x}_1 = \{x_{11}, \ldots, x_{1h}\}$ and $\forall \vec{x}_2 = \{x_{21}, \ldots, x_{2h}\}$. Let $\mathcal{K}^* = \partial_{\mathcal{K}_1}(\vec{y}_1)$ and $\mathcal{K}^* \not\models \bot$ such that $\mathcal{I}^p_{dm}(\mathcal{K}_1) = d^p_h(\vec{x}_1, \vec{y}_1)$. Since $\mathcal{K}_1 \overset{\triangle}{=} \mathcal{K}_2$ it follows that there exists $\vec{y}_2 = \{y_{21}, \ldots, y_{2h}\}$ with $y_{2i} = y_{1i}$ if $x_{2i} = x_{1i}$ or $y_{2i} = 1 - y_{1i}$ if $x_{2i} = 1 - x_{1i}$. Therefore, $\mathcal{I}^p_{dm}(\mathcal{K}_1) = d^p_{\bar{b}_{\mathcal{K}_1}}(\vec{x}_1, \vec{y}_1) = d^p_{\bar{b}_{\mathcal{K}_2}}(\vec{x}_2, \vec{y}_2) = \mathcal{I}^p_{dm}(\mathcal{K}_2)$.

(FCI). Let $\mathcal{K} = \{\kappa_1, \ldots, \kappa_h\}$, where $\kappa_i = c_i[\rho_i], \forall i = \overline{1, h}$. Let $\mathcal{I}^p_{dm}(\mathcal{K}) = \Delta$ and $\vec{x} = \{x_1, \ldots, x_h\}$ with $x_i \in [0, 1]$ such that $\partial_\mathcal{K}(\vec{x}) \not\models \bot$ and $|x_1 - \rho_1| + \cdots + |x_h - \rho_h| = \Delta$. Let $\mathcal{K}^* = \mathcal{K} \backslash \{\kappa^*\}$ with $\kappa^*[\rho^*] \in \mathsf{Fc}(\mathcal{K})$. It follows that $\vec{\rho}^* = \vec{\rho} \backslash \rho^*$ and $\vec{x}^* = \vec{x} \backslash x^*$. Since $\partial_\mathcal{K}(\vec{x}) \not\models \bot$, $|x^* - \rho^*| = 0$. Therefore, $\mathcal{I}^p_{dm}(\mathcal{K}) = \sqrt[p]{\sum_{i=1}^{h} |x_i - y_i|^p} = \sqrt[p]{\sum_{i=1}^{h-1} |x_i - y_i|^p + |x^* - \rho^*)|^p} = \mathcal{I}^p_{dm}(\mathcal{K}^*)$.

(SCI). As $\kappa \in \mathsf{Sc}(\mathcal{K})$, by Proposition 3 [53] $\kappa \in \mathsf{Fc}(\mathcal{K})$. Hence, $\mathcal{I}^p_{dm}(\mathcal{K}) = \mathcal{I}^p_{dm}(\mathcal{K} \backslash \{\kappa\})$.

(SUA). Let $\mathcal{K}_1 = \partial_\mathcal{K}(\vec{x}_1) \; \forall \vec{x}_1 = \{x_{11}, \ldots, x_{1\bar{b}_{\mathcal{K}_1}}\}$ and $\mathcal{K}_2 = \partial_\mathcal{K}(\vec{x}_2) \; \forall \vec{x}_2 = \{x_{21}, \ldots, x_{2\bar{b}_{\mathcal{K}_2}}\}$. Hence, there exists $\vec{y}_1, \vec{y}_2 \in [0,1]$ such that $\mathcal{K}_1 = \partial_{\mathcal{K}_1}(\vec{y}_1^*)$ and $\mathcal{K}_1 \not\models \bot$, $\mathcal{K}_2 = \partial_{\mathcal{K}_1}(\vec{y}_2^*)$ and $\mathcal{K}_2 \not\models \bot$. We have $\mathcal{I}^1_{dm}(\mathcal{K}_1 \cup \mathcal{K}_2) = d^1_{\bar{b}_{\mathcal{K}_1 \cup \mathcal{K}_2}}(\vec{x}_1 \cup \vec{x}_2, \vec{y}_1 \cup \vec{y}_2) = (|x_{11} - y_{11}| + \cdots + |x_{11} - y_{1\bar{b}_{\mathcal{K}_1}}|) + (|x_{21} - y_{21}| + \cdots + |x_{21} - y_{2\bar{b}_{\mathcal{K}_2}}|) = d^1_{\bar{b}_{\mathcal{K}_1}}(\vec{x}_1, \vec{y}_1) + d^1_{\bar{b}_{\mathcal{K}_2}}(\vec{x}_2, \vec{y}_2) = \mathcal{I}^1_{dm}(\mathcal{K}_1) + \mathcal{I}^1_{dm}(\mathcal{K}_2)$. □

Example 3.9 *Consider $\mathcal{K}_1$ and $\mathcal{K}_5$ in example 2.3. Consider a consistent PKB $\mathcal{K}[\vec{y}] = \langle (H)[0.8], (T)[0.9], (D)[0.5], (T|H)[1.0], (H|D)[1.0] \rangle$.*

By formula 3.20, $\vec{x} = \langle 0.7, 0.3, 0.45, 0.5, 0.64 \rangle$ and $\vec{y} = \langle 0.7, 0.4, 0.45, 0.5, 0.7 \rangle$.

By formula 3.20 and formula 3.1,

$\mathcal{I}^p_{dm}(\mathcal{K}_1) = \sqrt[p]{0.1^p + 0.6^p + 0.05^p + 0.5^p + 0.36^p}$.

Therefore, for $p = 1$, we have $\mathcal{I}^1_{dm}(\mathcal{K}_1) = 1.61$, for $p = 2$ we have $\mathcal{I}^2_{dm}(\mathcal{K}_1) = 0.87$.

Similarly, $\mathcal{I}^1_{dm}(\mathcal{K}_5) = 1.57$ and $\mathcal{I}^2_{dm}(\mathcal{K}_5) = 0.92$. For $p = 1$ we have $\mathcal{I}^1_{dm}(\mathcal{K}_1) + \mathcal{I}^1_{dm}(\mathcal{K}_5) = \mathcal{I}^1_{dm}(\mathcal{K}_1 \cup \mathcal{K}_5) = 3.18$.

*Therefore, $\mathcal{I}^1_{dm}$ of $\mathcal{K}_1$ and $\mathcal{K}_5$ fulfills **MIS** and **SUA** but fails to satisfy **NOR**.*

For $p = 2$, $\mathcal{I}^2_{dm}(\mathcal{K}_1) + \mathcal{I}^2_{dm}(\mathcal{K}_2) = 1.78 > \mathcal{I}^2_{dm}(\mathcal{K}_1 \cup \mathcal{K}_2) = 1.26$.

*Therefore, $\mathcal{I}^1_{dm}$ fails to satisfy **MIS**, and **SUA**.*

*Similarly in example 3.3, $\mathcal{I}^p_{dm}$ of $\mathcal{K}_1$ and $\mathcal{K}_5$ fails to satisfy **PEN**.*

Definition 3.27 *([7]) An inconsistency measure $\mathcal{I}^p_{sum} : \mathbb{K} \to \mathbb{R}^*$ of $\mathcal{K}$ is defined as follows:*

$$\mathcal{I}^p_{sum}(\mathcal{K}) = \sum_{\mathcal{M}^i \in \mathsf{SMI}(\mathcal{K})} \mathcal{I}^p_{dm}(\mathcal{M}^i) \tag{3.21}$$

Theorem 3.8 ([7]) $\mathcal{I}_{sum}^p$ *fulfills* **CON**, **MON**, **SUA**, **MIS**, **FCI**, **SCI**, *and* **PEN**.

Proof 3.8 *The proof of the above theorem is analogous to the proof for measures* $\mathcal{I}_{smi}^c$ *(Theorem 3.3) and* $\mathcal{I}_{dm}^p$ *(Theorem 3.7).* □

Example 3.10 *Consider* $\mathcal{K}_1$ *and* $\mathcal{K}_5$ *in example 2.3.*

By example 3.3, $\mathcal{M}^i[\vec{x}] = $ SMI$(\mathcal{K}_1) = \{\{(H)[0.7], (T)[0.3], (T|H)[0.5]\}\}$. *Consider consistent PKB* $\mathcal{M}^i[\vec{y}] = \langle (H)[0.8], (T)[0.9], (T|H)[1.0], \rangle$.

By formula 3.20 and formula 3.21, we have $\mathcal{I}_{sum}^p(\mathcal{K}_1) = \sqrt[p]{0.1^p + 0.6^p + 0.5^p}$. *For* $p = 1$, $\mathcal{I}_{sum}^p(\mathcal{K}_1) = 1.2$.

Therefore $\mathcal{I}_{sum}^p$ *of* $\mathcal{K}_1$ *fails to satisfy* **NOR**.

Similarly, $\mathcal{I}_{sum}^1(\mathcal{K}_5) = 1.17$. *For* $p = 2$, $\mathcal{I}_{sum}^p(\mathcal{K}_1) = 0.78$, $\mathcal{I}_{sum}^p(\mathcal{K}_5) = 0.83$.

Definition 3.28 ([7]) *Let* $\mathcal{K}$ *be a knowledge base; the probabilistic Shapley inconsistency value of* $\kappa \in \mathcal{K}$ *is defined as follows:*

$$SIV_{\mathcal{K}}(\kappa) = \sum_{\mathcal{K}_s \subseteq \mathcal{K}} \frac{(\bar{b}_{\mathcal{K}_s} - 1)!(\bar{b}_{\mathcal{K}} - \bar{b}_{\mathcal{K}_s})!}{\bar{b}_{\mathcal{K}}!} d \qquad (3.22)$$

where $d = \mathcal{I}_b(\mathcal{K}_s) - \mathcal{I}_b(\mathcal{K}_s \setminus \{\kappa\})$.

Definition 3.29 ([7]) *An inconsistency measure* $\mathcal{I}_{sv} : \mathbb{K} \to \mathbb{R}^*$ *of* $\mathcal{K}$ *is defined as follows:*

$$\mathcal{I}_{sv}(\mathcal{K}) = \max_{\kappa \in \mathcal{K}} \{SIV_{\mathcal{K}}(\kappa)\} \qquad (3.23)$$

Theorem 3.9 ([7]) $\mathcal{I}_{sv}$ *fulfills* **CON**, **NOR** *and* **FCI**.

Proof 3.9 *(CON). If* $\mathcal{K} \not\models \perp$, $\mathcal{I}_b(\mathcal{K}_s) - \mathcal{I}_b(\mathcal{K}_s \setminus \{\kappa\}) = 0 \forall \kappa \in \mathcal{K}$. *Hence,* $SIV_{\mathcal{K}}(\kappa) = 0 \forall \kappa \in \mathcal{K}$. *Therefore,* $\mathcal{I}_{sv}(\mathcal{K}) = 0$.

The proof of the properties **FCI** *and* **NOR** *is analogous to the proof of Proposition 4 in [7].* □

Example 3.11 *Consider* $\mathcal{K}_1$ *and* $\mathcal{K}_5$ *in example 2.3 with* $\mathcal{I}_{dr}(\mathcal{K}_1) = 1$, $\mathcal{I}_{dr}(\mathcal{K}_5) = 1$ *are basic inconsistency measures.*

Therefore, $SIV_{\mathcal{K}_1}((T)[0.7]) = SIV_{\mathcal{K}_1}((H)[0.7]) = SIV_{\mathcal{K}_1}((T)[0.3]) = SIV_{\mathcal{K}_1}((D)[0.45]) = SIV_{\mathcal{K}_1}((T|H)[0.5]), SIV_{\mathcal{K}_1}((H|D)[0.64]) = 4.8$.

Therefor, $\mathcal{I}_{sv}(\mathcal{K}_1) = 4.8$, *similarly* $\mathcal{I}_{sv}(\mathcal{K}_5) = 4.8$.

3.3.2   The Norm-based Inconsistency Measures

The minimal violation measures with respect to p-norm ($p > 1$ and $p \neq \infty$), 1-norm, and ∞-norm are introduced in Theorem 5.4, 5.15, 5.16 etc., respectively [40]. However, these measures are considered in the probabilistic-logic context. In this section, we also introduce several inconsistency measures for the probabilistic framework; that is, they are considered over E in $\mathcal{R}$ profile.

Deriving from Definition 3.7, we make Definition 3.30 to represent the inconsistency measure with respect to p-norm.

Definition 3.30 *[51] Let $\mathcal{R} = \langle \mathcal{B}, \mathsf{E} \rangle$ be PKB profile and $\mathcal{K} = \{(F_1 | G_1)[\rho_1], \ldots, \left(F_{\bar{b}_{\mathcal{K}}} \middle| G_{\bar{b}_{\mathcal{K}}}\right)[\rho_{\bar{b}_{\mathcal{K}}}]\}$ and $\vec{z} = (z_1, \ldots, z_{\bar{b}_{\mathcal{K}}})^T$, where $z_i = \mathcal{P}(F_i G_i) - \rho_i \mathcal{P}(G_i)$ with $(F_i | G_i)[\rho_i] \in \mathcal{K}$. The inconsistency measure with respect to p-norm ($p \geq 1$) of $\mathcal{K} \in \mathcal{B}$ over E is defined as follows:*

$$\mathcal{I}_{\mathsf{E}}^p(\mathcal{K}) = \min_{\vec{z} \in \mathbb{R}^{\bar{b}_{\mathcal{K}}}} \left\{ d^p(\mathcal{K}) \middle| A_{\mathcal{K}}^{\mathsf{E}} \mathcal{P} = \vec{z} \right\} \tag{3.24}$$

Theorem 3.10 is employed to calculate the inconsistency measure with respect to p-norm ($p > 1$ and $p \neq \infty$).

Theorem 3.10 *[51] Let $\mathcal{R} = \langle \mathcal{B}, \mathsf{E} \rangle$ be PKB profile and $f : \mathbb{R}^{\hbar_E} \to \mathbb{R}^*$ such that $f(\vec{\omega}) = \left\| A_{\mathcal{K}}^{\mathsf{E}} \cdot \vec{\omega} \right\|_p$. The inconsistency measure $\mathcal{I}_{\mathsf{E}}^p$ with respect to p-norm ($p > 1$ and $p \neq \infty$) of $\mathcal{K} \in \mathcal{B}$ over E is the optimal value f^* of the following optimization problem:*

$$\min_{\vec{\omega} \in \mathbb{R}^{\hbar_E}} \left\| A_{\mathcal{K}}^{\mathsf{E}} \cdot \vec{\omega} \right\|_p \tag{3.25}$$

subject to $\vec{\omega} \in C_p$
where, $C_p = \left\{ \vec{\omega} \in \mathbb{R}^{\hbar_E} \middle| \sum\limits_{i=1}^{\hbar_E} \omega_i = 1, \vec{\omega} \geq \vec{0} \right\}$.

Proof 3.10 *Because either $\mathsf{App}(\mathcal{K}) \cap \mathsf{E} = \varnothing$ or $\forall (F_i | G_i)[\rho_i] \in \mathcal{K}$, $\rho_i = 0$, where $i = \overline{1, h}$ makes $A_{\mathcal{K}}^{\mathsf{E}} = 0$, $\mathcal{K}$ is always consistent over E, in other word, $\mathcal{I}_{\mathsf{E}}^p(\mathcal{K}) = 0$ over E. If $\exists (F | G)[\rho] \in \mathcal{K}$ such that $\rho \neq 0$, we can prove that (a) there always exists a feasible solution for the problem (3.25) and (b) $\mathcal{I}_{\mathsf{E}}^p(\mathcal{K})$ is an inconsistency measure of $\mathcal{K}$.*

(a) Firstly, we can prove that the problem (3.25) is always feasible as follows: We have a constraint set C_p, and an objective function $f(\vec{\omega}) = \left\| A_{\mathcal{K}}^{\mathsf{E}} \cdot \vec{\omega} \right\|_p$, where $\omega_i \in \mathbb{R}^{\hbar_E}$.

Suppose that ϵ is the optimal value of the problem (3.25). Then there exists an optimal point $\vec{\omega}^*$, such that $\epsilon = \left\| A_{\mathcal{K}}^{\mathsf{E}} \cdot \vec{\omega}^* \right\|_p$. This leads to C_p, which is nonempty and thus $\vec{\omega}^*$ is feasible and can reach the optimal value g^*. Therefore, vector $\vec{\omega}^*$ is the optimal solution of the problem (3.25) [68].

(b) Secondly, we can prove that $\mathcal{I}_{\mathsf{E}}^p(\mathcal{K})$ is an inconsistency measure w.r.t p-norm of $\mathcal{K} \in \mathcal{B}$ over E. Truly, suppose that $\exists \vec{\omega}^* \in \mathbb{R}^{\hbar_{\mathsf{E}}}$ such that $\vec{\omega}^* \in C_p$. According to Definitions 3.3, 3.30, etc. and Formula with respect to p-norm, $\mathcal{I}_{\mathsf{E}}^p(\mathcal{K}) = \left\| A_{\mathcal{K}}^{\mathsf{E}} \cdot \vec{\omega}^* \right\|_p$. If $\vec{\omega}^* \in \mathbb{R}^{\hbar_{\mathsf{E}}}$ then $\vec{\omega}^*$ fulfills inequality constraints $\omega_i^* \geq 0 \ \forall i = \overline{1, \hbar_{\mathsf{E}}}$. This leads to $\mathcal{I}_{\mathsf{E}}^p(\mathcal{K}) \geq 0$.

We need to prove that $\mathcal{K}$ is consistent iff $\mathcal{I}_{\mathsf{E}}^p(\mathcal{K}) = 0$.

Necessity: Assume $\mathcal{K}$ is consistent, then there exists $\vec{\omega}^* \in \mathbb{R}^{\hbar_{\mathsf{E}}}$ such that $\left\| A_{\mathcal{K}}^{\mathsf{E}} \cdot \vec{\omega}^* \right\|_p = 0$, that is, $\mathcal{I}_{\mathsf{E}}^p(\mathcal{K}) = 0$.

Sufficiency: Assume that $\mathcal{I}_{\mathsf{E}}^p(\mathcal{K}) = 0$. By Definition 3.7, we have $\mho(\mathcal{K}) \neq \emptyset$. Therefore, $\mathcal{K}$ is consistent. $\qquad\square$

The following theorem is employed to calculate the inconsistency measure with respect to 1-norm.

Theorem 3.11 *[51] Let $\mathcal{R} = \langle \mathcal{B}, \mathsf{E} \rangle$ be PKB profile and $f : \mathbb{R}^{\hbar_{\mathsf{E}} + \bar{b}_{\mathcal{K}}} \to \mathbb{R}^*$ such that $f(\vec{\omega}, \vec{\lambda}) = \sum\limits_{i=1}^{\bar{b}_{\mathcal{K}}} \lambda_i$. The inconsistency measure $\mathcal{I}_{\mathsf{E}}^1$ with respect to 1-norm of $\mathcal{K} \in \mathcal{B}$ over E is the optimal value f^* of the following linear optimization problem:*

$$\min_{(\vec{\omega}, \vec{\lambda}) \in \mathbb{R}^{\hbar_{\mathsf{E}} + \bar{b}_{\mathcal{K}}}} \sum_{i=1}^{\bar{b}_{\mathcal{K}}} \lambda_i \tag{3.26}$$

subject to $(\vec{\omega}, \vec{\lambda}) \in C_1$
where, $C_1 = \left\{ (\vec{\omega}, \vec{\lambda}) \in \mathbb{R}^{\hbar_{\mathsf{E}} + \bar{b}_{\mathcal{K}}} \,\middle|\, A_{\mathcal{K}}^{\mathsf{E}} \cdot \vec{\omega} - \vec{\lambda} \leq \vec{0}, A_{\mathcal{K}}^{\mathsf{E}} \cdot \vec{\omega} + \vec{\lambda} \geq \vec{0}, \sum\limits_{i=1}^{\hbar_{\mathsf{E}}} \omega_i = 1, \vec{\omega} \geq \vec{0}, \vec{\lambda} \geq \vec{0} \right\}$.

Proof 3.11 *(a) Firstly, the problem (3.26) exists a feasible solution $(\vec{\omega}^*, \vec{\lambda}^*) \in C_1$ and reach the optimal value $\epsilon = \sum_{i=1}^{\bar{b}_{\mathcal{K}}} \lambda_i^*$. These proofs are similar to that of Theorem 3.10.*

(b) Secondly, we prove that $\mathcal{I}_{\mathsf{E}}^1(\mathcal{K})$ is an inconsistency measure with respect to 1-norm of $\mathcal{K} \in \mathcal{B}$ over E. Suppose that $\vec{\lambda}^* \in \mathbb{R}^{\bar{b}_{\mathcal{K}}}$ is a part of a feasible solution of the problem (3.26), that is, $\vec{\lambda}^* \in C_1$. According to Definitions 3.3, 3.30 etc., $\mathcal{I}_{\mathsf{E}}^1(\mathcal{K}) = \sum_{i=1}^{\bar{b}_{\mathcal{K}}} \lambda_i^*$. If $\vec{\lambda}^* \in \mathbb{R}^{\bar{b}_{\mathcal{K}}}$ then $\vec{\lambda}^*$ satisfies inequality constraints $\lambda_i^* \geq 0 \ \forall i = \overline{1, \bar{b}_{\mathcal{K}}}$. This leads to $\mathcal{I}_{\mathsf{E}}^1(\mathcal{K}) \geq 0$.

We need to prove that $\mathcal{K}$ is consistent iff $\mathcal{I}_{\mathsf{E}}^1(\mathcal{K}) = 0$.

Necessity: Assume $\mathcal{K}$ is consistent then there exists $\vec{\lambda}^* \in \mathbb{R}^{\bar{b}_{\mathcal{K}}}$ such that $\sum_{i=1}^{\bar{b}_{\mathcal{K}}} \lambda_i^* = 0$, that is, $\mathcal{I}_E^1(\mathcal{K}) = 0$.

Sufficiency: Assume that $\mathcal{I}_E^1(\mathcal{K}) = 0$. By Definition 3.7, we have $\mho(\mathcal{K}) \neq \emptyset$. Therefore, $\mathcal{K}$ is consistent. ☐

The following theorem is employed to calculate the inconsistency measure with respect to ∞-norm.

Theorem 3.12 *[51] Let $\mathcal{R} = \langle \mathcal{B}, \mathsf{E} \rangle$ be PKB profile and $f : \mathbb{R}^{\hbar_E + 1} \to \mathbb{R}^*$ such that $f(\vec{\omega}, \lambda) = \lambda$. The inconsistency measure $\mathcal{I}_E^\infty$ with respect to ∞-norm of $\mathcal{K} \in \mathcal{B}$ over E is the optimal value f^* of the linear optimization problem:*

$$\min_{(\vec{\omega}, \lambda) \in \mathbb{R}^{\hbar_E + 1}} \lambda \tag{3.27}$$

subject to $(\vec{\omega}, \lambda) \in C_\infty$
where, $C_\infty = \left\{ (\vec{\omega}, \lambda) \in \mathbb{R}^{\hbar_E + 1} \,\middle|\, A_{\mathcal{K}}^{\mathsf{E}} \cdot \vec{\omega} - \vec{1}\lambda \leq \vec{0}, A_{\mathcal{K}}^{\mathsf{E}} \cdot \vec{\omega} + \vec{1}\lambda \geq \vec{0}, \sum_{i=1}^{\hbar_E} \omega_i = 1, \vec{\omega} \geq \vec{0}, \lambda \geq 0 \right\}$

Proof 3.12 *(a) Firstly, the problem (3.27) exists a feasible solution $(\vec{\omega}^*, \lambda^*) \in C_\infty$ and reaches the optimal value $\epsilon = \lambda^*$. These proofs are similar to that of Theorem 3.10.*

(b) Secondly, we prove that $\mathcal{I}_E^\infty(\mathcal{K})$ is an inconsistency measure with respect to ∞-norm of $\mathcal{K} \in \mathcal{B}$ over E. Suppose that $\lambda^* \in \mathbb{R}^h$ is a part of feasible solution of the problem (3.27), that is, $\lambda^* \in C_\infty$. According to Definitions 3.3, 3.30 etc., $\mathcal{I}_E^\infty(\mathcal{K}) = \lambda^*$. If $\lambda^* \in \mathbb{R}^h$ then λ^* satisfies inequality constraints $\lambda^* \geq 0$. This leads to $\mathcal{I}_E^\infty(\mathcal{K}) \geq 0$.

We need to prove that $\mathcal{K}$ is consistent iff $\mathcal{I}_E^\infty(\mathcal{K}) = 0$.

Necessity: Assume that $\mathcal{K}$ is consistent, then there exists $\lambda^* \in \mathbb{R}$ such that $\lambda^* = 0$, that is, $\mathcal{I}_E^\infty(\mathcal{K}) = 0$.

Sufficiency: Assume that $\mathcal{I}_E^\infty(\mathcal{K}) = 0$. By Definition 3.7, we have $\mho(\mathcal{K}) \neq \emptyset$. Therefore, $\mathcal{K}$ is consistent. ☐

Theorem 3.13 is employed to present the relationship between inconsistency measures with respect to norm and desirable properties.

Theorem 3.13 *[69] The inconsistency measure $\mathcal{I}_p$ satisfies* **CON**, **MON**, **SUA**, **IRS**, *and* **SCI**.

Proof 3.13 *(CON). Assume that $\mathcal{K} \models \bot$.*

For $p = 1$, by Theorem 3.11 there exists $\vec{\lambda}^* \in \mathbb{R}^{\bar{b}_\mathcal{K}}$ such that $\sum_{i=1}^{\bar{b}_\mathcal{K}} \lambda_i^* = 0$. Hence, it follows $\mathcal{I}_E^1(\mathcal{K}) = 0$.

For $p = \infty$, by Theorem 3.12 there exists $\lambda^* \in \mathbb{R}$ such that $\lambda^* = 0$. Hence, it follows $\mathcal{I}_E^\infty(\mathcal{K}) = 0$.

For $p > 1(p \neq \infty)$, by Theorem 3.10 there exists $\vec{\omega}^* \in \mathbb{R}^{h_E}$ such that $\|A_\mathcal{K}^E \cdot \vec{\omega}^*\|_p = 0$. Hence, it follows $\mathcal{I}_E^p(\mathcal{K}) = 0$.

(MON). For $p = 1$, by Theorem 3.11 we have $\mathcal{I}_E^1(\mathcal{K} \cup \{\kappa\}) = \sum_{i=1}^{\bar{b}_\mathcal{K}+1} \lambda_i^* \geq \sum_{i=1}^{\bar{b}_\mathcal{K}} \lambda_i^* = \mathcal{I}_E^1(\mathcal{K})$.

For $p = \infty$, by Theorem 3.12 we have $\mathcal{I}_E^\infty(\mathcal{K} \cup \{\kappa\}) = \lambda^* + \lambda^{*'} \geq \lambda^* = \mathcal{I}_E^\infty(\mathcal{K})$.

For $p > 1(p \neq \infty)$, by Theorem 3.10, 3.3, etc., $\mathcal{I}_E^p(\mathcal{K} \cup \{\kappa\}) = \sqrt[p]{\sum_{i=1}^{\bar{b}_\mathcal{K}+1} |z_i|^p} \geq \sqrt[p]{\sum_{i=1}^{\bar{b}_\mathcal{K}} |z_i|^p} = \mathcal{I}_E^p(\mathcal{K})$.

(SUA). Let $\mathcal{K}_1 = \{\kappa_1, \ldots, \kappa_n\}$ and $\mathcal{K}_2 = \{\kappa_{n+1}, \ldots, \kappa_{n+m}\}$.

For $p = 1$, by Theorem 3.11 we have $\mathcal{I}_E^1(\mathcal{K}_1 \cup \mathcal{K}_2) = \sum_{i=1}^{n+m} \lambda_i^* = \sum_{i=1}^{n} \lambda_i^* + \sum_{i=n+1}^{n+m} \lambda_i^* = \mathcal{I}_E^1(\mathcal{K}_1) + \mathcal{I}_E^1(\mathcal{K}_2)$.

For $p = \infty$, by Theorem 3.12 we have $\mathcal{I}_E^\infty(\mathcal{K}_1 \cup \mathcal{K}_2) = \lambda^*$ and $\mathcal{I}_E^\infty(\mathcal{K}_1) + \mathcal{I}_E^\infty(\mathcal{K}_2) = \lambda^{*1} + \lambda^{*2}$. It is possible to $\lambda^{*1} + \lambda^{*2} \geq \lambda^*$. Therefore, this property could be violated.

For $p > 1(p \neq \infty)$, by Theorem 3.10, 3.3, etc., $\mathcal{I}_E^p(\mathcal{K}_1 \cup \mathcal{K}_2) = \sqrt[p]{\sum_{i=1}^{n+m} |z_i|^p} \geq \sqrt[p]{\sum_{i=1}^{n} |z_i|^p} + \sqrt[p]{\sum_{i=n+1}^{n+m} |z_i|^p} = \mathcal{I}_E^p(\mathcal{K}_1) + \mathcal{I}_E^p(\mathcal{K}_2)$.

(IRS). As $\mathcal{K}_1 \overset{\wedge}{=} \mathcal{K}_2$, due to Definition 3.17 it holds that there exists a bijection $\alpha : \mathcal{K}_1 \to \mathcal{K}_2$ such that $\kappa \equiv \alpha(\kappa) \ \forall \kappa \in \mathcal{K}_1$. Hence, $\bar{b}_{\mathcal{K}_1} = \bar{b}_{\mathcal{K}_2} = n$.

For $p = 1$, by Theorem 3.11 there exists $\vec{\lambda}^* = \{\lambda_1^*, \ldots, \lambda_n^*\}$ such that $\mathcal{I}_E^1(\mathcal{K}_1) = \sum_{i=1}^{n} \lambda_i^*$ and $\mathcal{I}_E^1(\mathcal{K}_2) = \sum_{i=1}^{n} \lambda_i^*$. Therefore, $\mathcal{I}_E^1(\mathcal{K}_1) = \mathcal{I}_E^1(\mathcal{K}_2)$.

For $p = \infty$, by Theorem 3.12 there exists $\lambda^* \in \mathbb{R}$ such that $\mathcal{I}_E^\infty(\mathcal{K}_1) = \lambda^*$ and $\mathcal{I}_E^\infty(\mathcal{K}_2) = \lambda^*$. Therefore, $\mathcal{I}_E^\infty(\mathcal{K}_1) = \mathcal{I}_E^\infty(\mathcal{K}_2)$.

For $p > 1(p \neq \infty)$, by Theorem 3.10 $A_{\mathcal{K}_1}^E = A_{\mathcal{K}_2}^E \ \forall \vec{\omega} \in \mathbb{R}$. Thus $\|A_{\mathcal{K}_1}^E \cdot \vec{\omega}\|_p = \|A_{\mathcal{K}_2}^E \cdot \vec{\omega}\|_p$. Therefore, $\mathcal{I}_E^p(\mathcal{K}_1) = \mathcal{I}_E^p(\mathcal{K}_2)$.

(SCI). Let $\mathcal{K}_1 = \mathcal{K}\backslash\{\kappa\}$ and $\mathcal{K}_2 = \{\kappa\}$. Let $z^* = \mathcal{P}(FG) - \rho\mathcal{P}(G)$ correspond to $\kappa = (F|G)[\rho]$ with $\kappa \in \mathcal{K}_2$. Since $\kappa \in \text{Sc}(\mathcal{K})$, by Definition 3.12 it follows that $\kappa \in \mathcal{K}$ such that $\text{App}(\kappa) \cap \text{App}(\mathcal{K}\backslash\{\kappa\}) = \emptyset$. Hence, $\mathcal{K}_2 \models \bot$. It follows that

$z^* = 0$. *By Definition 3.3,* $\mathcal{I}_E^p(\mathcal{K}) = \min_{\vec{\omega} \in \mathbb{R}^{n_E}} \left\| A_\mathcal{K}^E \cdot \vec{\omega} \right\|_p \leq \sqrt[p]{\sum_{i=1}^{b_\mathcal{K}-1} |z_i|^p + |z^*|^p} =$

$\sqrt[p]{\sum_{i=1}^{b_\mathcal{K}-1} |z_i|^p + 0} = \mathcal{I}_E^p(\mathcal{K}_1) = \mathcal{I}_E^p(\mathcal{K} \setminus \{\kappa\})$. *Since* $\mathcal{I}_E^p(\mathcal{K})$ *fulfills the property* **MON**, $\mathcal{I}_E^p(\mathcal{K})$
$\geq \mathcal{I}_E^p(\mathcal{K} \setminus \{\kappa\})$. *Therefore,* $\mathcal{I}_E^p(\mathcal{K}) = \mathcal{I}_E^p(\mathcal{K} \setminus \{\kappa\})$. □

Example 3.12 *Let's continue example 3.2, we compute* $\mathcal{I}_E^1(\mathcal{K}_1)$, $\mathcal{I}_E^2(\mathcal{K}_1)$, *and* $\mathcal{I}_E^\infty(\mathcal{K}_1)$.
  - By theorem 3.11, $\mathcal{I}_E^1(\mathcal{K}_1)$ *is computed by solving the following linear optimization problem:*

$$\min_{(\vec{\omega}, \vec{\lambda}) \in \mathbb{R}^{8+5}} (\lambda_1 + \lambda_2 + \lambda_3 + \lambda_4 + \lambda_5) \tag{3.28}$$

subject to:

$$0.3\omega_1 + 0.3\omega_2 + 0.3\omega_3 + 0.3\omega_4 - 0.7\omega_5 - 0.7\omega_6 - 0.7\omega_7 - 0.7\omega_8 - \lambda_1 \leq 0 \tag{3.29}$$

$$0.7\omega_1 + 0.7\omega_2 - 0.3\omega_3 - 0.3\omega_4 + 0.7\omega_5 + 0.7\omega_6 - 0.3\omega_7 - 0.3\omega_8 - \lambda_2 \leq 0 \tag{3.30}$$

$$0.55(\omega_1 + \omega_3 + \omega_5 + \omega_7) - 0.45(\omega_2 + \omega_4 + \omega_6 + \omega_8) - \lambda_3 \leq 0 \tag{3.31}$$

$$0.5\omega_1 + 0.5\omega_2 - 0.5\omega_3 - 0.5\omega_4 - \lambda_4 \leq 0 \tag{3.32}$$

$$0.36\omega_1 + 0.36\omega_3 - 0.64\omega_5 - 0.64\omega_7 - \lambda_5 \leq 0 \tag{3.33}$$

$$0.3\omega_1 + 0.3\omega_2 + 0.3\omega_3 + 0.3\omega_4 - 0.7\omega_5 - 0.7\omega_6 - 0.7\omega_7 - 0.7\omega_8 + \lambda_1 \geq 0 \tag{3.34}$$

$$0.7\omega_1 + 0.7\omega_2 - 0.3\omega_3 - 0.3\omega_4 + 0.7\omega_5 + 0.7\omega_6 - 0.3\omega_7 - 0.3\omega_8 + \lambda_2 \geq 0 \tag{3.35}$$

$$0.55(\omega_1 + \omega_3 + \omega_5 + \omega_7) - 0.45(\omega_2 + \omega_4 + \omega_6 + \omega_8) + \lambda_3 \geq 0 \tag{3.36}$$

$$0.5\omega_1 + 0.5\omega_2 - 0.5\omega_3 - 0.5\omega_4 + \lambda_4 \geq 0 \tag{3.37}$$

$$0.36\omega_1 + 0.36\omega_3 - 0.64\omega_5 - 0.64\omega_7 + \lambda_5 \geq 0 \tag{3.38}$$

$$\omega_1 + \omega_2 + \omega_3 + \omega_4 + \omega_5 + \omega_6 + \omega_7 + \omega_8 = 1 \tag{3.39}$$

$$\omega_1 \geq 0, \omega_2 \geq 0, \omega_3 \geq 0, \omega_4 \geq 0, \omega_5 \geq 0, \omega_6 \geq 0, \omega_7 \geq 0, \omega_8 \geq 0 \tag{3.40}$$

$$\lambda_1 \geq 0, \lambda_2 \geq 0, \lambda_3 \geq 0, \lambda_4 \geq 0, \lambda_5 \geq 0 \tag{3.41}$$

An optimal solution of problem (3.28) is $(\vec{\omega}^*, \vec{\lambda}^*)$,
where $\vec{\omega}^* = (0, 0.35, 0.29, 0.06, 0, 0, 0.16, 0.14)$ *and* $\vec{\lambda}^* = (0, 0.05, 0, 0)$. *Therefore,*
$\mathcal{I}_E^1(\mathcal{K}_1) = 0 + 0.05 + 0 + 0 = 0.5$.
  - By theorem 3.10, $\mathcal{I}_E^2(\mathcal{K}_1)$ *is computed by solving the following non-linear optimization problem:*

$$\min_{\vec{\omega} \in \mathbb{R}^8} \sqrt{|t_1|^2 + |t_2|^2 + |t_3|^2 + |t_4|^2 + |t_5|^2} \tag{3.42}$$

subject to:

$$\omega_1 + \omega_2 + \omega_3 + \omega_4 + \omega_5 + \omega_6 + \omega_7 + \omega_8 = 1 \tag{3.43}$$

$$\omega_1 \geq 0, \omega_2 \geq 0, \omega_3 \geq 0, \omega_4 \geq 0, \omega_5 \geq 0, \omega_6 \geq 0, \omega_7 \geq 0, \omega_8 \geq 0 \tag{3.44}$$

where, $t_1 = 0.3\omega_1 + 0.3\omega_2 + 0.3\omega_3 + 0.3\omega_4 - 0.7\omega_5 - 0.7\omega_6 - 0.7\omega_7 - 0.7\omega_8$, $t_2 = 0.7\omega_1 + 0.7\omega_2 - 0.3\omega_3 - 0.3\omega_4 + 0.7\omega_5 + 0.7\omega_6 - 0.3\omega_7 - 0.3\omega_8$, $t_3 = 0.55(\omega_1 + \omega_3 + \omega_5 + \omega_7) - 0.45(\omega_2 + \omega_4 + \omega_6 + \omega_8)$, $t_4 = 0.5\omega_1 + 0.5\omega_2 - 0.5\omega_3 - 0.5\omega_4$, $t_5 = 0.36\omega_1 + 0.36\omega_3 - 0.64\omega_5 - 0.64\omega_7$.

An optimal solution of problem (3.42) is $\vec{\omega}^* = (0.15, 0, 0, 0.32, 0.29, 0.16, 0, 0.08)$. *Therefore,* $\mathcal{I}_E^2(\mathcal{K}_1) = 0.33$.

- By theorem 3.12 , $\mathcal{I}_E^\infty(\mathcal{K}_1)$ *is computed by solving the following linear optimization problem:*

$$\min_{(\vec{\omega}, \lambda) \in \mathbb{R}^{8+1}} \lambda \tag{3.45}$$

subject to:

$$0.3\omega_1 + 0.3\omega_2 + 0.3\omega_3 + 0.3\omega_4 - 0.7\omega_5 - 0.7\omega_6 - 0.7\omega_7 - 0.7\omega_8 - \lambda \leq 0 \tag{3.46}$$

$$0.7\omega_1 + 0.7\omega_2 - 0.3\omega_3 - 0.3\omega_4 + 0.7\omega_5 + 0.7\omega_6 - 0.3\omega_7 - 0.3\omega_8 - \lambda \leq 0 \tag{3.47}$$

$$0.55(\omega_1 + \omega_3 + \omega_5 + \omega_7) - 0.45(\omega_2 + \omega_4 + \omega_6 + \omega_8) - \lambda \leq 0 \tag{3.48}$$

$$0.5\omega_1 + 0.5\omega_2 - 0.5\omega_3 - 0.5\omega_4 - \lambda \leq 0 \tag{3.49}$$

$$0.36\omega_1 + 0.36\omega_3 - 0.64\omega_5 - 0.64\omega_7 - \lambda \leq 0 \tag{3.50}$$

$$0.3\omega_1 + 0.3\omega_2 + 0.3\omega_3 + 0.3\omega_4 - 0.7\omega_5 - 0.7\omega_6 - 0.7\omega_7 - 0.7\omega_8 + \lambda \geq 0 \tag{3.51}$$

$$0.7\omega_1 + 0.7\omega_2 - 0.3\omega_3 - 0.3\omega_4 + 0.7\omega_5 + 0.7\omega_6 - 0.3\omega_7 - 0.3\omega_8 + \lambda \geq 0 \tag{3.52}$$

$$0.55(\omega_1 + \omega_3 + \omega_5 + \omega_7) - 0.45(\omega_2 + \omega_4 + \omega_6 + \omega_8) + \lambda \geq 0 \tag{3.53}$$

$$0.5\omega_1 + 0.5\omega_2 - 0.5\omega_3 - 0.5\omega_4 + \lambda \geq 0 \tag{3.54}$$

$$0.36\omega_1 + 0.36\omega_3 - 0.64\omega_5 - 0.64\omega_7 + \lambda \geq 0 \tag{3.55}$$

$$\omega_1 + \omega_2 + \omega_3 + \omega_4 + \omega_5 + \omega_6 + \omega_7 + \omega_8 = 1 \tag{3.56}$$

$$\omega_1 \geq 0, \omega_2 \geq 0, \omega_3 \geq 0, \omega_4 \geq 0, \omega_5 \geq 0, \omega_6 \geq 0, \omega_7 \geq 0, \omega_8 \geq 0 \tag{3.57}$$

$$\lambda \geq 0 \tag{3.58}$$

An optimal solution of problem (3.45) is $(\vec{\omega}^*, \vec{\lambda}^*)$, *where* $\vec{\omega}^* = (0, 0.32, 0.32, 0.04, 0, 0, 0.15, 0.17)$ *and* $\lambda^* = 0.02$.

Therefore, $\mathcal{I}_E^\infty(\mathcal{K}_1) = 0.02$. *This result is suitable for the statement that* $\mathcal{K}_1$ *is inconsistent according to Example 3.1.*

3.3.3 The Unnormalized Inconsistency Measure

Another approach in suggesting inconsistency measures is not based on the *p*-norm, and is called unnormalized inconsistency measure. The measure $\mathsf{Inc}^+(\mathcal{K})$ [65,

38] is an inconsistency measure based on the probability value of some probabilistic constraints in the knowledge base that causes inconsistencies.

However, the measure $\mathsf{Inc}^+(\mathcal{K})$ is considered in the propositional language context. This section introduces the unnormalized inconsistency measure of a PKB extended from $\mathsf{Inc}^+(\mathcal{K})$ [65, 38]. The unnormalized inconsistency measure is considered in the probabilistic context, that is, they are considered over E in $\mathcal{R}$ profile.

We make Definition 3.31 to compute an unnormalized inconsistency measure.

Definition 3.31 *[54, 70] Let $\mathcal{R} = \langle \mathcal{B}, \mathsf{E} \rangle$ be PKB profile and $f : \mathbb{R}^{\hbar_E + 2\bar{b}_\mathcal{K}} \to \mathbb{R}^*$ such that $f(\vec{\omega}, \vec{\Delta}) = \sum_{i=1}^{\bar{b}_\mathcal{K}} (\ell_i + \zeta_i)$. An unnormalized inconsistency measure $\mathcal{I}_\mathsf{E}^u$ of $\mathcal{K} \in \mathcal{B}$ over E is the optimal solution of the non-linear optimization problem:*

$$\min_{(\vec{\omega}, \vec{\Delta}) \in \mathbb{R}^{\hbar_E + 2\bar{b}_\mathcal{K}}} \sum_{i=1}^{\bar{b}_\mathcal{K}} (\ell_i + \zeta_i) \tag{3.59}$$

subject to: $(\vec{\omega}, \vec{\Delta}) \in C_u$
where, $\vec{\Delta} = (\vec{\ell}, \vec{\zeta}) = (\ell_1, \ldots, \ell_{\bar{b}_\mathcal{K}}, \zeta_1, \ldots, \zeta_{\bar{b}_\mathcal{K}})$,
$\Upsilon = C_\mathcal{K}^{E,+} \vec{\omega} + (\vec{\rho} + \vec{\ell} - \vec{\zeta}) C_\mathcal{K}^{E,-} \vec{\omega}$ and
$C_u = \left\{ (\vec{\omega}, \vec{\Delta}) \in \mathbb{R}^{\hbar_E + 2\bar{b}_\mathcal{K}} | \bar{A}_\mathcal{K} \cdot \vec{\Delta}^\top \leq \vec{1} - \vec{\rho}, \bar{A}_\mathcal{K} \cdot \vec{\Delta}^\top \geq -\vec{\rho}, \sum_{i=1}^{\hbar_E} \omega_i = 1, \vec{\omega} \geq \vec{0}, \Upsilon = 0 \right\}$
is the set of constraints for the problem with computing an unnormalized inconsistency measure.

Theorem 3.14 is employed to calculate the unnormalized inconsistency measure.

Theorem 3.14 *Let $\mathcal{R} = \langle \mathcal{B}, \mathsf{E} \rangle$ be PKB profile and $\mathcal{K} \in \mathcal{B}$. Measure $\mathcal{I}_\mathsf{E}^u$ is an inconsistency measure of $\mathcal{K}$.*

Proof 3.14 *The proof is similar to that of Theorem 3.10.*
(a) Problem (3.59) exists a feasible solution $(\vec{\omega}^*, \vec{\Delta}^*) \in C_u$ with $\vec{\Delta}^* = (\vec{\ell}^*, \vec{\zeta}^*)$ and reach the optimal value $\epsilon = \sum_{i=1}^{\bar{b}_\mathcal{K}} (\ell_i^* + \zeta_i^*)$.
(b) Secondly, we prove that $\mathcal{I}_\mathsf{E}^u(\mathcal{K})$ is an unnormalized inconsistency measure of $\mathcal{K} \in \mathcal{B}$ over E. Suppose that $\vec{\Delta}^* \in \mathbb{R}^{2\bar{b}_\mathcal{K}}$ is a part of feasible solution of problem (3.59), that is, $\vec{\Delta}^* \in C_u$. Therefore, $\mathcal{I}_\mathsf{E}^u(\mathcal{K}) = \sum_{i=1}^{\bar{b}_\mathcal{K}} (\ell_i^* + \zeta_i^*)$.

If $\vec{\Delta}^* \in \mathbb{R}^{2\bar{b}_\mathcal{K}}$ then by Definition of the problem (3.59), $\sum_{i=1}^{\bar{b}_\mathcal{K}} (\ell_i^* + \zeta_i^*) \geq 0$. This leads to $\mathcal{I}_\mathsf{E}^u(\mathcal{K}) \geq 0$.

We need to prove that $\mathcal{K}$ is consistent iff $\mathcal{I}_\mathsf{E}^u(\mathcal{K}) = 0$.

Necessity: Assume $\mathcal{K}$ is consistent, then there exists $\vec{\Delta}^* = (\vec{\ell}^*, \vec{\zeta}^*) \in \mathbb{R}^{2\bar{b}_{\mathcal{K}}}$ such that $\sum_{i=1}^{\bar{b}_{\mathcal{K}}} (\ell_i^* + \zeta_i^*) = 0$, that is, $\mathcal{I}_{\mathsf{E}}^u(\mathcal{K}) = 0$.

Sufficiency: Assume that $\mathcal{I}_{\mathsf{E}}^u(\mathcal{K}) = 0$. By Definition 3.7, we have $\mho(\mathcal{K}) \neq \emptyset$. Therefore, $\mathcal{K}$ is consistent. □

Theorem 3.15 is employed to present the relationship between the unnormalized inconsistency measures and desirable properties.

Theorem 3.15 *[69] The inconsistency measure $\mathcal{I}^u$ satisfies **CON**, **MON**, **SUA**, **IRS**, **FCI**, and **SCI**.*

Proof 3.15 *(**CON**). Assume that $\mathcal{K} \models \bot$. By Definition 3.31 there exists $\vec{\Delta} = (\vec{\ell}, \vec{\zeta}) = (\ell_1, \ldots, \ell_{\bar{b}_{\mathcal{K}}}, \zeta_1, \ldots, \zeta_{\bar{b}_{\mathcal{K}}})$ such that $\sum_{i=1}^{\bar{b}_{\mathcal{K}}} (\ell_i + \zeta_i) = 0$.*

Hence, it follows $\mathcal{I}_{\mathsf{E}}^u(\mathcal{K}) = 0$.

*(**MON**). By Definition 3.31, we have $\mathcal{I}_{\mathsf{E}}^u(\mathcal{K} \cup \{\kappa\}) = \sum_{i=1}^{\bar{b}_{\mathcal{K}}+1} (\ell_i^* + \zeta_i^*) \geq \sum_{i=1}^{\bar{b}_{\mathcal{K}}} \ell_i^* + \zeta_i^* = \mathcal{I}_{\mathsf{E}}^u(\mathcal{K})$.*

*(**SUA**). Let $\mathcal{K}_1 = \{\kappa_1, \ldots, \kappa_n\}$ and $\mathcal{K}_2 = \{\kappa_{n+1}, \ldots, \kappa_{n+m}\}$.*

By Definition 3.31, we have $\mathcal{I}_{\mathsf{E}}^u(\mathcal{K}_1 \cup \mathcal{K}_2) = \sum_{i=1}^{n+m} (\ell_i + \zeta_i)$, $\mathcal{I}_{\mathsf{E}}^u(\mathcal{K}_1) = \sum_{i=1}^{n} (\ell_i + \zeta_i)$ and $\mathcal{I}_{\mathsf{E}}^u(\mathcal{K}_2) = \sum_{i=n+1}^{m} (\ell_i + \zeta_i)$.

Therefore, $\mathcal{I}_{\mathsf{E}}^u(\mathcal{K}_1 \cup \mathcal{K}_2) = \sum_{i=1}^{n+m} (\ell_i + \zeta_i) = \sum_{i=1}^{n} (\ell_i + \zeta_i) + \sum_{i=n+1}^{m} (\ell_i + \zeta_i) = \mathcal{I}_{\mathsf{E}}^u(\mathcal{K}_1) + \mathcal{I}_{\mathsf{E}}^u(\mathcal{K}_2)$.

*(**IRS**). As $\mathcal{K}_1 \stackrel{\wedge}{=} \mathcal{K}_2$, due to Definition 3.17 it holds that there exist a bijection $\alpha : \mathcal{K}_1 \to \mathcal{K}_2$ such that $\kappa \equiv \alpha(\kappa) \ \forall \kappa \in \mathcal{K}_1$. Hence, $\bar{b}_{\mathcal{K}_1} = \bar{b}_{\mathcal{K}_2} = n$.*

By Definition 3.31, there exists $\vec{\Delta} = (\vec{\ell}, \vec{\zeta}) = (\ell_1, \ldots, \ell_{\bar{b}_{\mathcal{K}}}, \zeta_1, \ldots, \zeta_{\bar{b}_{\mathcal{K}}})$, such that $\mathcal{I}_{\mathsf{E}}^u(\mathcal{K}_1) = \sum_{i=1}^{n} (\ell_i + \zeta_i)$ and $\mathcal{I}_{\mathsf{E}}^u(\mathcal{K}_2) = \sum_{i=1}^{n} (\ell_i + \zeta_i)$.

Therefore, $\mathcal{I}_{\mathsf{E}}^u(\mathcal{K}_1) = \mathcal{I}_{\mathsf{E}}^u(\mathcal{K}_2)$.

*(**SCI**). Let $\mathcal{K}_1 = \mathcal{K} \backslash \{\kappa\}$ and $\mathcal{K}_2 = \{\kappa\}$. Since $\kappa \in \mathsf{Sc}(\mathcal{K})$, by Definition 3.12 it follows that $\kappa \in \mathcal{K}$ such that $\mathsf{App}(\kappa) \cap \mathsf{App}(\mathcal{K} \backslash \{\kappa\}) = \emptyset$. Hence, $\mathcal{K}_2 \not\models \bot$. It follows that there exists ℓ^*, ζ^* such that $\mathcal{I}_{\mathsf{E}}^u(\mathcal{K}_2) = \ell^* + \zeta^* = 0$.*

By Definition 3.31, $\mathcal{I}_{\mathsf{E}}^u(\mathcal{K}) = min_{(\vec{\omega}, \vec{\Delta}) \in \mathbb{R}^{h_{\mathsf{E}} + 2\bar{b}_{\mathcal{K}}}} \sum_{i=1}^{\bar{b}_{\mathcal{K}}} (\ell_i + \zeta_i) \leq \sum_{i=1}^{\bar{b}_{\mathcal{K}}-1} (\ell_i + \zeta_i) + \ell^* + \zeta^* = \mathcal{I}_{\mathsf{E}}^u(\mathcal{K}_1) = \mathcal{I}_{\mathsf{E}}^u(\mathcal{K} \backslash \{\kappa\})$.

*Since $\mathcal{I}_{\mathsf{E}}^u(\mathcal{K})$ fulfills the property **MON**, $\mathcal{I}_{\mathsf{E}}^u(\mathcal{K}) \geq \mathcal{I}_{\mathsf{E}}^u(\mathcal{K} \backslash \{\kappa\})$. Therefore, $\mathcal{I}_{\mathsf{E}}^u(\mathcal{K}) = \mathcal{I}_{\mathsf{E}}^u(\mathcal{K} \backslash \{\kappa\})$.* □

Example 3.13 *Let's continue example 3.2. By Definition 3.31, $\mathcal{I}_E^u(\mathcal{K}_1)$ is computed by solving the following non-linear optimization problem:*

$$\min_{(\vec{\omega},\vec{\Delta})\in\mathbb{R}^{8+10}} (\ell_1 + \ell_2 + \ell_3 + \ell_4 + \ell_5 + \zeta_1 + \zeta_2 + \zeta_3 + \zeta_4 + \zeta_5) \tag{3.60}$$

subject to:

$$\ell_1 - \zeta_1 \le 0.3, \ell_2 - \zeta_2 \le 0.7, \ell_3 - \zeta_3 \le 0.55, \ell_4 - \zeta_4 \le 0.5, \ell_5 - \zeta_5 \le 0.36 \tag{3.61}$$

$$\ell_1 + 0.7 \ge \zeta_1, \ell_2 + 0.3 \ge \zeta_2, \ell_3 + 0.45 \ge \zeta_3, \ell_5 + 0.5 \ge \zeta_4, \ell_5 + 0.64 \ge \zeta_5 \tag{3.62}$$

$$(\omega_1 + \omega_2 + \omega_3 + \omega_4) - (0.7 + \ell_1 - \zeta_1)\sum_{i=1}^{8}\omega_i = 0 \tag{3.63}$$

$$(\omega_1 + \omega_2 + \omega_5 + \omega_6) - (0.3 + \ell_2 - \zeta_2)\sum_{i=1}^{8}\omega_i = 0 \tag{3.64}$$

$$(\omega_1 + \omega_3 + \omega_5 + \omega_7) - (0.45 + \ell_3 - \zeta_3)\sum_{i=1}^{8}\omega_i = 0 \tag{3.65}$$

$$(\omega_1 + \omega_2) - (0.5 + \ell_4 - \zeta_4)(\omega_1 + \omega_2 + \omega_3 + \omega_4) = 0 \tag{3.66}$$

$$(\omega_1 + \omega_3) - (0.64 + \ell_5 - \zeta_5)(\omega_1 + \omega_3 + \omega_5 + \omega_7) = 0 \tag{3.67}$$

$$\omega_1 + \omega_2 + \omega_3 + \omega_4 + \omega_5 + \omega_6 + \omega_7 + \omega_8 = 1 \tag{3.68}$$

$$\omega_1 \ge 0, \omega_2 \ge 0, \omega_3 \ge 0, \omega_4 \ge 0, \omega_5 \ge 0, \omega_6 \ge 0, \omega_7 \ge 0, \omega_8 \ge 0 \tag{3.69}$$

An optimal solution of problem (3.60) is $(\vec{\omega}^*, \vec{\Delta}^*)$,
where
$\vec{\Delta}^* = (\vec{\ell}^*, \vec{\zeta}^*)$ with
$\vec{\ell}^* = (0, 0.05, 0, 0, 0)$,
$\vec{\zeta}^* = (0, 0, 0, 0, 0)$.
Therefore, $\mathcal{I}_E^u(\mathcal{K}_1) = 0.05$.

Table 3.2 summarizes the basic, norm-based and unnormalized inconsistency measures from the above examples.

Table 3.2 The inconsistency measures of PKBs $\mathcal{K}_1, \mathcal{K}_2, \mathcal{K}_3, \mathcal{K}_4$ and $\mathcal{K}_5$

PKBs	$\mathcal{I}_{dr}$	$\mathcal{I}_{mi}$	$\mathcal{I}_{smi}^c$	$\mathcal{I}_\ell$	$\mathcal{I}_\chi$	$\mathcal{I}_\mu$	$\mathcal{I}_{dm}^1$	$\mathcal{I}_{dm}^2$
$\mathcal{K}_1$	1	1	0.33	0.5	7	1	1.61	0.87
$\mathcal{K}_2$	0	0	0	0	0	0	0	0
$\mathcal{K}_3$	0	0	0	0	0	0	0	0
$\mathcal{K}_4$	0	0	0	0	0	0	0	0
$\mathcal{K}_5$	1	1	0.33	0.5	7	1	1.57	0.92

PKBs	$\mathcal{I}_{sum}^1$	$\mathcal{I}_{sum}^2$	$\mathcal{I}_{sv}$	$\mathcal{I}_E^1(\mathcal{K})$	$\mathcal{I}_E^2(\mathcal{K})$	$\mathcal{I}_E^\infty(\mathcal{K})$	$\mathcal{I}_E^u(\mathcal{K})$
$\mathcal{K}_1$	1.2	0.78	4.8	0.05	0.033	0.02	0.05
$\mathcal{K}_2$	0	0	0	0	0	0	0
$\mathcal{K}_3$	0	0	0	0	0	0	0
$\mathcal{K}_4$	0	0	0	0	0	0	0
$\mathcal{K}_5$	1.17	0.83	4.8	0.026	0.018	0.011	0.026

Table 3.3 summarizes the relationship between the inconsistency measures and desirable properties

Table 3.3 The relationship between the inconsistency measures and desirable properties

Properties	$\mathcal{I}_{dr}$	$\mathcal{I}_{mi}$	$\mathcal{I}_{smi}^c$	$\mathcal{I}_\ell$	$\mathcal{I}_\chi$	$\mathcal{I}_\mu$	$\mathcal{I}_{dm}^1$	$\mathcal{I}_{dm}^p$	$\mathcal{I}_{sum}^p$	$\mathcal{I}_{sv}$	$\mathcal{I}_\mathcal{K}^p$	$\mathcal{I}_\mathcal{K}^u$
CON	√	√	√	√	√	√	√	√	√	√	√	√
MON	√	√	√	√	√	√	√	√	√	-	√	√
SUA	-	√	√	-	-	√	√	-	√	√	√	√
NOR	√	-	-	√	-	-	-	-	-	-	-	-
MIS	-	√	√	-	-	√	-	-	√	-	-	-
FCI	√	√	√	√	√	√	√	√	√	√	-	-
SCI	√	√	√	√	√	√	√	√	√	-	√	√
IRS	√	√	-	√	√	-	√	√	-	-	√	√
PEN	-	√	√	-	-	√	-	-	√	-	√	√
Theorems	3.1	3.2	3.3	3.4	3.5	3.6	3.7	3.7	3.8	3.9	3.13	3.15

3.4 ALGORITHMS FOR COMPUTING THE INCONSISTENCY MEASURES

3.4.1 The Computational Complexity

To build a model for solving the consistency restoring problem and the problems with integrating PKBs, it is necessary to solve linear optimization problems and convex optimization problems. These problems could be solved by simplex method [68, 71] or interior-point method [56]. Therefore, in order to evaluate the complexity of algorithms, the book will first present the complexity of the interior-point method to solve the following convex optimization problem:

$$\min f_0(x) \tag{3.70}$$

subject to

$$f_i(x) \leq 0, i = \overline{1, \hbar} \tag{3.71}$$

$$Qx = b \tag{3.72}$$

One of these methods is the barrier method. Now assume that the problem (3.70) is solvable. Let $\hbar$ be the number of inequality constraints. Let $\mu = 1 + \frac{1}{\sqrt{\hbar}}$, ϵ be a guaranteed specified accuracy, $c = \log_2(\log_2\frac{1}{\epsilon})$. Let function $\Phi(x) = -\sum\limits_{i=1}^{\hbar} \log(-f_i(x))$ be the logarithmic barrier for problem (3.70). Therefore, the gradient of $\Phi(x)$ is

$$\nabla\Phi(x) = \sum_{i=1}^{\hbar} \frac{1}{-f_i(x)} \nabla f_i(x) \tag{3.73}$$

$t^{(0)}$ is the value that minimizes

$$\inf_{\mathcal{G}} \|t\nabla f_0(x^{(0)}) + \nabla\Phi(x^{(0)}) + Q^T \mathcal{G}\|_2 \tag{3.74}$$

The constant $\gamma = \frac{\alpha\beta(1-2\alpha)}{20-8\alpha}$ depends on the backtracking parameters α and β. Then, the complexity of the interior-point method for solving problem (3.70) is

$$N = \frac{\log \frac{\hbar}{t^{(0)}\epsilon}}{\log \mu} \left(\frac{\hbar(\mu - 1 - \log\mu)}{\gamma} + c \right) \tag{3.75}$$

3.4.2 The General Methods

Before presenting the algorithms in this book, the common methods for algorithms will be introduced as follows: - The problem type is determined by using method `OptimizationProblem`(*Optype*), where *Optype* is the input parameter. If *Optype*='*MINIMIZE*', the minimization problem will be solved, and If *Optype*='*MAXIMIZE*', the maximization problem will be solved. All algorithms in this book use the parameter '*MINIMIZE*'.

- A set of complete conjunctions of events could be found by employing method `FindingSSC`(*Ev*), where *Ev* is a set of events **E**.

- An expression could be added into a constraint or an objective function by employing method `addCVar`($\langle CVar \rangle$), where *CVar* is an expression.

- A constraint could be added into a set of constraints by employing method addCs($\langle Cs \rangle$), where Cs is an inequality constraint $(g_i(x) \leq 0, g_i(x) \geq 0)$ or an equality constraint $(h_j(x) = 0)$.

- An objective function could be made by employing method addOF($\langle Obf \rangle$), where Obf is the linear or non-linear function.

- Optimization problems could be solved by employing method OpenOpt.solve(Prt, Pr), where Pr is a problem, and Prt is the problem type including linear programming (LP) and non-linear programming (NLP). This method uses $OpenOp$ package. An LP problem has the objective function and all of the constraints that are linear. The NLP problem has several constraints or the objective function are non-linear.

- Quadratic problems (QP) could be solved by employing method OpenojAlgo.solve(Prt, Pr), where Pr is a problem, and Prt is the quadratic problem. This method uses $ojAlgo$ package. A QP problem has an objective function that is quadratic and constraints that are linear.

- The optimal solution of the problem could be obtained by employing method getOptimalSolution(sol), where sol is the result of OpenOpt.solve(Prt, Pr) or OpenojAlgo.solve(Prt, Pr).

- The optimal value of the problem could be obtained by employing method getOptimalValue(sol), where sol is the result of OpenOpt.solve(Prt, Pr) or OpenojAlgo.solve(Prt, Pr).

- The probability value of constraints could be obtained by employing method getValue($Cons$), where $Cons$ is a probabilistic constraint.

- The initial object could be initialized by employing method setEmpty(). DDFs that was introduced in Tables 5.2 and 5.3 could be built by employing method BuildDF(X, y, d), where X is a matrix including n-rows in which each row is the satisfying probability vectors of a PKB in PKB profile, y is a probability merging vector and d is the form of DDF.

Details of the code for these methods can be found in an online appendix (https://thamnguyenvan.blogspot.com/).

3.4.3 Algorithms

The following algorithm 1 (**NBOP**-*p-Norm Based Optimization Problems*) is employed to build and solve optimization problems with respect to norms.

1) Algorithm **NBOP**:

Input: A PKB, a set of events, and p-norm.

Output: The solution of optimization problem.

Idea: For each norm, the objective function and constraints are constructed. For 1-norm and ∞-norm, the optimal solution of the linear problem could be found by using the simplex method. For 2-norm, the optimal solution of the non-linear optimization problem could be found by using the interior-point method.

Algorithm 1 Building and solving optimization problems with respect to norms(NBOP) [51]

Input : $\langle \mathcal{K}, \mathsf{E}, p \rangle$
Output: sol

```
 1  Function NBOP(K, E, p)
    begin
 2      pr = OptimizationProblem(MINIMIZE); Fcc = FindingSCC(E);
 3      sω.setEmpty();
 4      for cc ∈ Fcc do
 5          |  pr.addCs(⟨ω[cc] ≥ 0⟩); sω.addCVar(⟨ω[cc]⟩);
 6      end for
 7      pr.addCs(⟨sω = 1⟩); g.setEmpty();
 8      for c[ρ] ∈ K do
 9          lie.setEmpty();
10          for cc ∈ Fcc do
11              |  if p=1 or p=∞ then
12              |      if App(c[ρ]) ⊆ App(cc) then lie.addCVar(⟨(1 − ρ) · ω[cc]⟩);
13              |      else lie.addCVar(⟨−ρ · ω[cc]⟩);
14              |  end if
15              |  if p =2 then
16              |      if App(c[ρ]) ⊆ App(cc) then g.addCVar(⟨(1 − ρ)²(ω[cc])²⟩);
17              |      else g.addCVar(⟨(−ρ)²(ω[cc])²⟩);
18              |  end if
19          end for
20          if p =1 then
21              |  pr.addCs(⟨λ[c] ≥ 0⟩); pr.addCs(⟨lie − λ[c] ≤ 0⟩); pr.addCs(⟨lie + λ[c] ≥ 0⟩);
22              |  g.addCVar(⟨λ[c]⟩);
23          end if
24          if p =∞ then
25              |  pr.addCs(⟨λ ≥ 0⟩); pr.addCs(⟨lie − λ ≤ 0⟩); pr.addCs(⟨lie + λ ≥ 0⟩);
26              |  g.addCVar(⟨λ⟩);
27          end if
28      end for
29      if p=1 or p=∞ then  sol = OpenOpt.solve(LP, pr.addOF(⟨g⟩));
30      if p=2 then  sol = OpenojAlgo.solve(QP, pr.addOF(⟨√g⟩));
31      return sol
32  end
```

Algorithm 1 (**NBOP**) builds and solves optimization problems with respect to 1-norm, 2-norm, and ∞-norm by using Theorem 3.11 and Theorem 5.3, Theorem 3.10 and Theorem 5.2, and Theorem 3.12 and Theorem 5.4, respectively.

Algorithm 1 consists of two stages,

(i) Building the optimization problem (from line 2 to 28).

- From line 3 to 7, building constraints $\sum_{i=1}^{\hbar_E} \omega_i = 1, \vec{\omega} \geq \vec{0}$.

- For 1-norm:

+ From line 11 to 14 and from line 20 to 23, building constraints $A_{\mathcal{K}}^E \cdot \vec{\omega} - \vec{\lambda} \leq \vec{0}$, $A_{\mathcal{K}}^E \cdot \vec{\omega} + \vec{\lambda} \geq \vec{0}, \vec{\lambda} \geq \vec{0}$;

+ Line 22 builds the objective function $g(\vec{\omega}, \vec{\lambda}) = \sum_{i=1}^{\bar{b}_{\mathcal{K}}} \lambda_i$.

- For 2-norm: From line 15 to 18, building the objective function $g(\vec{\omega}) = \left\| A_{\mathcal{K}}^E \vec{\omega} \right\|_2$.

- For ∞-norm:

+ From line 11 to 14 and from line 24 to 27, building constraints $A_{\mathcal{K}}^E \cdot \vec{\omega} - \vec{1}\lambda \leq \vec{0}, A_{\mathcal{K}}^E \cdot \vec{\omega} + \vec{1}\lambda \geq \vec{0}, \lambda \geq 0$;

+ Line 26 builds the objective function $g(\vec{\omega}, \vec{\lambda}) = \lambda$.

(ii) Solving the optimization problem (from line 29 to 30).

Algorithm 2 (**UNOP**-*Unnormalized Optimization Problems*) is employed to build and solve unnormalized optimization problems.

2) Algorithm UNOP.

Input: A PKB, a set of events.

Output: The solution of unnormalized optimization problem.

Idea: Firstly, the objective function and constraints are constructed. Secondly, the optimal solution of the unnormalized optimization problem could be found by using the interior-point method.

Algorithm 2 Building and solving unnormalized optimization problem (UNOP)

Input : $\langle \mathcal{K}, \mathsf{E} \rangle$
Output: *sol*

```
1  Function UNOP(𝒦, E)
   begin
2  │   pr = OptimizationProblem(MINIMIZE); Fcc = FindingSCC(E);
3  │   sω.setEmpty();
4  │   for  cc ∈ Fcc do
5  │   │   pr.addCs(⟨ω[cc] ≥ 0⟩); sω.addCVar(⟨ω[cc]⟩);
6  │   end for
7  │   pr.addCs(⟨sω = 1⟩);
8  │   g.setEmpty();
9  │   for  c[ρ] ∈ 𝒦 do
10 │   │   g.addCVar(⟨ℓ[c] + ζ[c]⟩);
11 │   │   pr.addCs(⟨ℓ[c] − ζ[c] ≤ 1 − ρ⟩); pr.addCs(⟨ℓ[c] − ζ[c] ≥ −ρ⟩);
12 │   end for
13 │   for  c[ρ] ∈ 𝒦 do
14 │   │   lie.setEmpty(); rie.setEmpty();
15 │   │   for  cc ∈ Fcc do
16 │   │   │   if App(c[ρ]) ⊆ App(cc) then lie.addCVar(⟨ω[cc]⟩);
17 │   │   │   if Right(c) ⊆ App(cc) then rie.addCVar(⟨ω[cc]⟩);
18 │   │   end for
19 │   │   pr.addCs(⟨(ρ + ℓ[c] − ζ[c])lei − rei = 0⟩);
20 │   end for
21 │   sol = OpenOpt.solve(NLP, pr.addOF(⟨g⟩));
22 │   return sol
23 end
```

Algorithm 2 is based on Definition 3.31 and Theorem 3.14, and consists of two stages:

(i) Building the optimization problem (from line 2 to 20):

- From line 3 to 7, building constraints $\sum_{i=1}^{\hbar_{\mathsf{E}}} \omega_i = 1, \vec{\omega} \geq \vec{0}$.

- From line 8 to 12, building constraints $\bar{A}_{\mathcal{K}} \cdot \vec{\Delta}^\top \leq \vec{1} - \vec{\rho}, \bar{A}_{\mathcal{K}} \cdot \vec{\Delta}^\top \geq -\vec{\rho}$.

- Line 10 builds the objective function $\sum_{i=1}^{\bar{b}_{\mathcal{K}}} \ell_i + \sum_{i=1}^{\bar{b}_{\mathcal{K}}} \zeta_i$.

- From line 13 to 20, building constraint $\Upsilon = 0$.

(ii) Solving the optimization problem (line 21).

Algorithm 3 (**CIM**-*Computing Inconsistency Measure*) is employed to compute inconsistency measures by using Algorithm **NBOP** and Algorithm **UNOP**.

3) Algorithm CIM:

Input: A PKB, a set of events, and *pu* to identify the problem type

Output: Inconsistency measure of a PKB over a set of events.

Idea: Depending on the input *pu*, the algorithm calls the optimal solution of the corresponding optimization problem. The optimal value of the optimization problem is calculated, and it is the inconsistency measure.

Algorithm 3 Computing inconsistency measure (CIM)

Input : $\langle \mathcal{K}, \mathsf{E}, pu \rangle$
Output: $\mathcal{I}(\mathcal{K})$

1 Function CIM($\mathcal{K}, \mathsf{E}, pu$)
  begin
2 | if $pu=1$ or $pu=2$ or $pu=\infty$ then $sol = \text{NBOP}(\mathcal{K}, \mathsf{E}, pu)$;
3 | if $pu=u$ then $sol = \text{UNOP}(\mathcal{K}, \mathsf{E})$;
4 | $\mathcal{I}(\mathcal{K}) = \text{getOptimalValue}(sol)$;
5 | return $\mathcal{I}(\mathcal{K})$
6 end

Algorithm 3 is based on Theorem 3.10, Theorem 3.11, Theorem 3.12, and Theorem 3.14; it consists of two stages:

(i) Call the optimal solution of the optimization problem with respect to pu (from line 2 to 3).

- If $pu = 1$ or $pu = 2$ or $pu = \infty$, using Algorithm NBOP(line 2).

- If $pu = u$, using Algorithm UNOP (line 3).

(ii) Call the method to get the optimal value corresponding to the inconsistency measure of the input PKB (line 4).

By Theorem 3.11 and Theorem 3.12, inconsistency measures with respect to 1-norm, ∞-norm is the solution of the linear optimization problem. Therefore, the class of linear problems could be solved by simplex method [68, 71]. By Theorem 3.10, the inconsistency measure with respect to 2-norm is the solution of the convex optimization problem. Therefore, the convex optimization problem can be solved by the interior-point method [56].

Theorem 3.16 is employed to evaluate the complexity of Algorithm 3 (**CIM**) for calculating inconsistency measures.

Theorem 3.16 *[51] Let $\mathcal{R} = \langle \mathcal{B}, \mathsf{E} \rangle$ be the PKB profile and $\mathcal{K} \in \mathcal{B}$. The complexity of Algorithm 3 (**CIM**) for calculating inconsistency measures of $\mathcal{K}$ is:*
  - $\mathcal{O}(\bar{b}_\mathcal{K} \times 3^{\bar{b}_\mathcal{K}})$ with respect to 1-norm, ∞-norm
  - $\mathcal{O}(\bar{b}_\mathcal{K}^2 \times 3^{\bar{b}_\mathcal{K}})$ with respect to p-norm ($p > 1$ and $p \neq \infty$) and unnormalized form.

Proof 3.16 *For 1-norm: In stage 1, by Definition 3.15 and Theorem 3.11, the objective function and constraints are linear so the cost for building the optimization problem is $\mathcal{O}\left(\bar{b}_\mathcal{K} \times \hbar_\mathsf{E}\right)$. In stage 2, the inconsistency measure $\mathcal{I}^1(\mathcal{K})$ is the optimal value of the problem (3.26) that is solved by simplex method [68, 71]. Thus,* the cost of the stage 2 is $\mathcal{O}\left(\bar{b}_{\mathcal{K}}^3 + \bar{b}_{\mathcal{K}} \times \hbar_{\mathsf{E}}\right)$. Therefore, the complexity of Algorithm 3 (**CIM**) for computing inconsistency measures with respect to 1-norm is $\mathcal{O}\left(\max\left\{\bar{b}_{\mathcal{K}}^3 + \bar{b}_{\mathcal{K}} \times \hbar_{\mathsf{E}}, \bar{b}_{\mathcal{K}} \times \hbar_{\mathsf{E}}\right\}\right) = \mathcal{O}\left(\bar{b}_{\mathcal{K}}^3 + \bar{b}_{\mathcal{K}} \times \hbar_{\mathsf{E}}\right)$. Let $n = \bar{b}_{\mathcal{K}}$ and $g(n) = n^2 + n2^n$. The fact that $g(n)$ is $\mathcal{O}(n \times 3^n)$ means $\mathcal{O}\left(\bar{b}_{\mathcal{K}}^3 + \bar{b}_{\mathcal{K}} \times \hbar_{\mathsf{E}}\right) = \mathcal{O}(\bar{b}_{\mathcal{K}} \times 3^{\bar{b}_{\mathcal{K}}})$.

Similarly, the complexity of Algorithm 3 (**CIM**) for computing inconsistency measures with respect to ∞-norm is $\mathcal{O}(\bar{b}_{\mathcal{K}} \times 3^{\bar{b}_{\mathcal{K}}})$.

For p-norm: In stage 1, by Definition 3.15 and Theorem 3.10, constraints are linear so the cost for building the optimization problem is $\mathcal{O}\left(\bar{b}_{\mathcal{K}} \times \hbar_{\mathsf{E}}\right)$. In stage 2, by Theorem 3.10, the objective function is non-linear. The inconsistency measure $\mathcal{I}^p(\mathcal{K})$ is the optimal value of the problem (3.25) that is solved by interior-point method [56]. By problem (3.70) with constraints (3.71–3.72) and by the problem (3.25), $f_0(\vec{\omega}) = \|A_{\mathcal{K}}^{\mathsf{E}} \vec{\omega}\|_p$, $f_i(\vec{\omega}) = -\omega_i \forall i = \overline{1, \hbar_{\mathsf{E}}}$, $Q = (\underbrace{1, \ldots, 1}_{\hbar_{\mathsf{E}}})$. By problem (3.73),

$$\nabla \Phi(\vec{\omega}) = \sum_{i=1}^{\hbar_{\mathsf{E}}} \frac{1}{\omega_i} \text{ so } t_{NIM}^{(0)} \text{ is the minimized value } \inf_{\mathcal{G}} \|t \nabla f_0(\vec{\omega}^{(0)}) + \nabla \Phi(\vec{\omega}^{(0)}) + Q^T \mathcal{G}\|_2.$$

Let N_p be the cost for solving the optimization problem (3.25). Utilizing problem (3.75) and (4.4), $N_p = \frac{\log \frac{\hbar_{\mathsf{E}}}{t_{NIM}^{(0)} \varepsilon}}{\log \mu} \left(\frac{\hbar_{\mathsf{E}}(\mu - 1 - \log \mu)}{\gamma} + c\right)$ so the cost of the stage 2 is $\mathcal{O}(N_p)$. Therefore, the complexity of Algorithm 3 (**CIM**) for computing the inconsistency measure with respect to p-norm is $\mathcal{O}\left(max\left\{\bar{b}_{\mathcal{K}} \times \hbar_{\mathsf{E}}, N_p\right\}\right)$. Let $g(n) = N_p$, we have $g(n) = \frac{n - \log t}{\log(2^{\frac{n}{2}} + 1) - \frac{n}{2}}\left(\frac{2^{\frac{n}{2}}}{\gamma} - \frac{2^n \log(2^{\frac{n}{2}} + 1)}{\gamma} + \frac{n2^n}{2\gamma} + c\right)$ where t, γ, c are constants. The fact that $g(n)$ is $\mathcal{O}(n^2 3^n)$. Therefore, $\mathcal{O}\left(max\left\{\bar{b}_{\mathcal{K}} \times \hbar_{\mathsf{E}}, N_p\right\}\right) = \mathcal{O}(\bar{b}_{\mathcal{K}}^2 \times 3^{\bar{b}_{\mathcal{K}}})$.

The unnormalized inconsistency measure: By Definition 3.31 and Theorem 3.14, the objective function is linear; however constraints are non-linear. Thus, the cost for building the optimization problem is $\mathcal{O}\left(\bar{b}_{\mathcal{K}} \times \hbar_{\mathsf{E}}\right)$. In stage 2, the inconsistency measure $\mathcal{I}_{\mathcal{K}}^u$ is the optimal value of the problem (3.59) that is solved by interior-point method [56]. By problem (3.70) with constraints (3.71–3.72) and by problem (4.24), $f_0(\vec{\omega}) = \sum_{i=1}^{\bar{b}_{\mathcal{K}}} \ell_i + \sum_{i=1}^{\bar{b}_{\mathcal{K}}} \zeta_i$, $f_i(\vec{\omega}) = -\omega_i \forall i = \overline{1, \hbar_{\mathsf{E}}}$, $Q = (\underbrace{1, \ldots, 1}_{\hbar_{\mathsf{E}}})$. By (3.73), $\nabla \Phi(x) = \sum_{i=1}^{3 \times h + m} \frac{1}{-f_i(\vec{\omega})} \nabla f_i(\vec{\omega})$ so $t_{UIM}^{(0)}$ is the minimized value $\inf_{\mathcal{G}} \|t \nabla f_0(\vec{\omega}^{(0)}) + \nabla \Phi(\vec{\omega}^{(0)}) + Q^T \mathcal{G}\|_2$. Let N_u be the cost for solving optimization problem (3.59). Utilizing problem (3.75) and (4.24), $N_u = \frac{\log \frac{3 \times \bar{b}_{\mathcal{K}} + \hbar_{\mathsf{E}}}{t_{UIM}^{(0)} \varepsilon}}{\log \mu} \left(\frac{(3 \times \bar{b}_{\mathcal{K}} + \hbar_{\mathsf{E}})(\mu - 1 - \log \mu)}{\gamma} + c\right)$ so the cost of stage 2 is $\mathcal{O}(N_{UIM})$. The complexity of Algorithm 3 (**CIM**) for calculating the unnormalized inconsistency measures is $\mathcal{O}\left(max\left\{\bar{b}_{\mathcal{K}} \times \hbar_{\mathsf{E}}, N_u\right\}\right)$. Let $g(n) = N_u$, we have $g(n) = \frac{\log(3n + 2^n) - \log t}{\log(1 + \frac{1}{\sqrt{3n + 2^n}})}(\frac{1}{\gamma}(3n +$

$2^n)(\frac{1}{\sqrt{3n+2^n}} - \log(1 + \frac{1}{\sqrt{3n+2^n}})) + c)$ *where* t, γ, c *are constants. The fact that* $g(n)$ *is* $\mathcal{O}(n^2 \times 3^n)$. *Therefore,* $\mathcal{O}\left(max\left\{\bar{b}_\mathcal{K} \times \hbar_E, N_u\right\}\right) = \mathcal{O}(\bar{b}_\mathcal{K}^2 \times 3^{\bar{b}_\mathcal{K}})$. □

3.5 CONCLUDING REMARKS

In this chapter, we have compiled the most common inconsistency measures for a logical, probabilistic-logic framework and adapted them to the probabilistic framework. We then offer a family of a desirable properties, as well as the assessment of desirable properties for the inconsistency measure of a PKB. Moreover, we make a deep survey on how to calculate inconsistency measures for the probabilistic framework. Moreover, we explored a deep survey on how to calculate inconsistency measures for the probabilistic framework. Particularly, we build algorithms for computing inconsistency measures with respect to norm and the unnormalized inconsistency measure. These algorithms are implemented by solving optimization problems. We also present the assessment of the complexity of algorithms through the proof of theorems. In the next chapter, we deal with processing inconsistency of the PKBs.

CHAPTER 4

Methods for restoring consistency in probabilistic knowledge bases

IN THIS CHAPTER, we first present a model to change an inconsistent PKB into a consistent one. We then discuss several techniques for restoring consistency in PKBs. Finally, we present several consistency-restoring algorithms and analyze and discuss their complexity.

4.1 OVERVIEW OF HANDLING INCONSISTENCIES

4.1.1 The Inconsistency Resolution Problem

In knowledge-based systems, the concept of inconsistency of knowledge is most often understood as a feature of knowledge characterized by the lack of possibility for inference processes. The sources of inconsistency can be classified regarding two aspects: centralization aspect and distribution aspect [1].

In terms of centralization aspect, there could appear inconsistencies within each knowledge base itself [72]. The reasons of inconsistency are:

(i) the newly acquired knowledge is inconsistent with the original knowledge because of a change in the real world to which it refers, for example in a multi-agent system;

(ii) the indeterminacy of relationships between events happening in the real world;

(iii) the inaccurate knowledge of acquisition, reception and processing devices;

(iv) the knowledge-processing procedures are not identifiable;

(v) the extracted knowledge depends on the data source and some of its components may be inconsistent.

In terms of distribution aspect, inconsistencies between different knowledge bases [73] could occur. The reasons for inconsistency are:

(i) Knowledge conflict occurs when several experts share their knowledge to solve a problem; since each expert has his or her own views, this can lead to an inconsistent general knowledge base;

(ii) the system performs the actual tasks by using a variety of knowledge sources that are usually processed autonomously and which employ different mechanisms to transfer the knowledge bases to the same representing form;

(iii) the uncertainty and incompleteness of the knowledge could induce inconsistencies;

(iv) the indeterminacy of knowledge-processing mechanisms in a knowledge-based system.

For the centralization aspect, from the integrity point of view, there is no benefit in an inconsistent knowledge base. An inconsistent knowledge base cannot be in use. Therefore, inconsistencies needs to be eliminated so that the knowledge integration systems work correctly to achieve the best common knowledge. However, for the distribution aspect, the inconsistency of knowledge is a tangibility which should not be eliminated but need to be solved for achieving the worthy knowledge. It is difficult to identify inconsistency in a knowledge base, as well as addressing the inconsistency. However, if inconsistencies are not resolved, the knowledge management will not be possible.

In the process of integrating knowledge bases, inconsistencies of knowledge can appear in two main stages: pre- and post-execution processing stages. Before the integration process is performed, inconsistency may occur in the initial knowledge itself. After the integration process is performed, inconsistency occurs in the resulting knowledge. Thus if the resulting knowledge base is inconsistent, it is very difficult to guarantee the reliability of this knowledge. Inconsistencies can also appear at intermediate stages of the integration process.

If the resulting knowledge base is inconsistent, it is very difficult to guarantee the reliability of the knowledge. Although only a small portion of the knowledge base is influenced by those inconsistencies, the whole system that uses this knowledge base can be seriously affected. Therefore, ensuring consistency of knowledge-based systems is always one of the essential requirements because, without it, most of these systems become useless [1, 74, 75]. Following this reason, many studies have involved the restoration of consistency in knowledge systems.

The inconsistency resolution problem for probabilistic knowledge-based systems is defined as follows:

Given an inconsistent PKB $\mathcal{K}$ over E, it is necessary to determine a consistent PKB $\mathcal{K}^*$ which best represents the given PKB.

4.1.2   Methods of Handling Inconsistencies

In general, it is difficult problem to identify inconsistencies in a knowledge base and to deal with them. However, maintaining consistency of the knowledge base is a vital issue in the design of knowedge-based systems [1, 74]. In order to solve inconsistency of a knowledge base, there are some developed strategies such as removing formulas [3, 4, 5, 6, 40], adjusting formulas [11, 12], and changing probability [6, 11, 34, 38, 40, 57, 65, 66, 67, 76]. Table 4.1 summarizes the techniques to address inconsistency of knowledge bases.

Formula-removing method:

Let $\mathcal{K}$ be a PKB, the formula elimination method is an approach to find $\mathcal{K}^*$ such that $\mathcal{K}^* \subseteq \mathcal{K}$ and $\mathcal{K}^*$ is consistent.

Hunter and Konieczny [7, 8, 9] suggest that the minimal inconsistent subsets are the cause of occuring inconsistencies in the knowledge base. Therefore, removing several formulas from the minimal inconsistent subsets will obtain a consistent knowledge base. In a logical knowledge base, inconsistency is addressed by removing the smallest set of formulas from the original knowledge base so that the remaining set is consistent [3, 4, 5]. Potyka [40] proposes a consistency-restoring operator $\Gamma : \mathbb{K} \to \mathbb{K}$ based on the kernel-restoring operator [5]. The idea of using the kernel operator of $\mathcal{K}$ is to consider the consistency-restoring process as the discarding of at least one element from each minimal inconsistent set of the probabilistic-logic knowledge base. Finthammer et al. [6] propose an approach by removing a minimum number of rules from the initial PKB $\mathcal{K}$ until the consistency is restored. However, the method of removing formulas from PKB can cause the loss of useful rules.

Table 4.1 Methods for solving inconsistency in knowledge bases

Methods	Idea	Knowledge base	Disadvantage
Removing formulas	Based on minimal inconsistent sets	Logic [3, 4, 5] Probabilistic-logic [40] Probability[6]	Losing some useful rules
Changing formulas	Adding further constraints	Probability [11, 12]	The theorems have not been proved Causing other inconsistencies
Changing Probability	Minimizing inconsistencies	Heuristic model Probability [6]	Depends on experience
		Probabilistic-logic[40, 66, 76]	New probability value finding difficulty
		Probabilistic-logic[57]	Only use probability bounds
		Probability [11]	The theorems have not been proved
		Probability [38, 65, 67]	Optimal value finding difficulty
		Probability [34]	Candidacy finding difficulty

Formula-altering method:

Let $\mathcal{K}$ be a PKB $\mathcal{K} = \{(F_1 | G_1) [\rho_1], \ldots, (F_n | G_n) [\rho_n]\}$, the formula altering method is an approach to find $\mathcal{K}^* = \{(F_1^* | G_1^*) [\rho_1], \ldots, (F_n^* | G_n^*) [\rho_n]\}$ so that $\mathcal{K}^*$ is consistent.

Rödder and Xu [11] find a consistent knowledge base $\mathcal{K}^*$ that is qualitatively the same as inconsistent knowledge base $\mathcal{K}$ and is closest to $\mathcal{K}$, that is, only modifying the qualitative structure of the knowledge base without altering the value of ρ_i. Each formula $(F_i | G_i) [\rho_i] \in \mathcal{K}$ [11] is extended to $(F_i | G_i H_i) [\rho_i]$ with a new proposition $H_i \, \forall i = \overline{1, n}$. Kern-Isberner and Rödder [12] improve the approach in [11] by extending $(F_i | G_i) [\rho_i]$ to $(F_i | G_i H) [\rho_i] \forall i = \overline{1, n}$ with a new proposition H. However, Kern and Rödder do not give justification and evaluation of any of these approaches. On the other hand, adding events to a probabilistic constraint can cause the self-contradictory probabilistic constraint or the conflict with other probabilistic constraints in the knowledge base if the relation of this probability constraint to others or to the basic principles of probability are not considered and evaluated carefully.

Probability-changing method:

Let $\mathcal{K}$ be a PKB $\mathcal{K} = \{(F_1 | G_1) [\rho_1], \ldots, (F_n | G_n) [\rho_n]\}$, the probability-changing method is an approach to find $\mathcal{K}^* = \{(F_1 | G_1) [\rho_1^*], \ldots, (F_n | G_n) [\rho_n^*]\}$ such that $\mathcal{K}^*$ is consistent.

In the approach of Finthammer et al. [6], the new probability value ρ^* of each probabilistic constraint $(F | G) \in \mathcal{K}$ has to satisfy the condition $\rho^* \in [d(F|G), t(F|G)]$, where $d(F|G)$ and $t(F|G)$ are lower and upper bounds for $(F|G)$. However, the approach of Finthammer et al. [6] has to obey a heuristic model and be guided by the knowledge engineer to determine the importance of constraints in PKB. Based on the determined significance level, the model will gradually change the probability through experience until consistency is restored.

Rödder and Xu [11] introduces two methods according to this approach. The idea of the first method is to gradually reduce the respective probabilities $\rho_1, \ldots, \rho_h$ corresponding to the constraints in $\mathcal{K}$ until $\mathcal{K}$ becomes consistent PKB. The second method uses general divergence as a distance measure to determine new probabilities, that is, the log-ratio $\log\left(\frac{\mathcal{P}(\bar{F}G)}{\mathcal{P}(FG)} \frac{\rho}{1-\rho}\right)$ is minimized to find new probabilities. However, there is no justification and evaluation of this approach given in [11].

Thimm [38, 65] uses the inconsistency measure $\mathcal{I}^*(\mathcal{K})$ to handle inconsistencies. The variables $y_i \in [0,1]\ \forall i = \overline{1,h}$ are employed to calculate the deviation of the values of the probabilistic constraints $(F_i) | G_i) [\rho_i]$ from consistent ones in a minimal way. Each constraint $(F_i | G_i) [\rho_i]$ is modified to $(F_i | G_i) [\rho_i + y_i^*]$, where $\vec{y}^* \overset{def}{=} \arg\min_{\vec{y} \in \mathbb{R}^h} \sum_{i=1}^h (|y_i|)$. Picado-Muino [67] also uses a class of the inconsistency measures to build a model for assessing inconsistencies in knowledge bases. Picado-Muino is also interested in adjusting minimum degrees of certainty (i.e. probability) to make $\mathcal{K}$ consistent. However, it is difficult for these approaches to compute in practice because they correspond to non-convex optimization problems suffering from the existence of non-global local minima.

Daniel [34] solves the problem of restoring consistency by exploiting the strengths of the principle of maximum entropy *(MEe)*. Daniel proposes a paraconsistent maximum entropy inference process *(inference process)* $\mathcal{I}_{ME}^E \overset{def}{=} \arg\max_{w \in \widehat{\Omega}_C} - \sum_{i=1}^m w_i \log w_i$ to solve the problem of electing a unique probability distribution that best represents consistent knowledge base. However, Daniel does not explain in detail how to find the best candidacy function for the model and the complexity of the optimization problem has not been considered and discussed in detail.

Potyka, Thimm [40, 66, 76] and Bona [57] employed minimal violation measures with respect to the p-norm $\mathcal{I}^p$ to restore the consistency of a PKB. Potyka and Thimm [76] propose a consistency-restoring operator Γ_p^{ME} for $\mathcal{K} = \{\mathcal{P}(F_i | G_i) = q_i | 1 \leq i \leq$

n}. Bona [57] propose a consistency-restoring operator Γ_p^ε for $\mathcal{K} = \{\mathcal{P}(F_i|G_i) \geq q_i | 1 \leq i \leq n\}$. Potyka [40, 66] deals with inconsistencies by employing minimal violation p-inconsistency measure with integrity constraints $\mathcal{I}_{IC}^p$ to build a consistency-restoring operator Γ_{IC}^p. The consistency-recovery method [40, 76] is similar to the second method proposed by Rödder and Xu [11] but a new probability value will be found by minimizing $|\mathcal{P}(FG) - \rho\mathcal{P}(G)|$. The approaches of Potyka, Thimm and Bona based on the principle of maximum entropy has overcome the limitation of the heuristic model proposed by Finthammer [6]. However, this solution is considered and applied to probabilistic-logic knowledge bases. It is difficult to find new probability values from probability functions satisfying the PKB because it is necessary to make sure the probability function satisfies all the probabilistic constraints in the PKB.

4.2 RESTORING CONSISTENCY IN PROBABILISTIC KNOWLEDGE BASES

4.2.1 Basic Notions

First of all, the thesis presents the concepts used as the basis for building a consistency recovery model and knowledge integration model. The following definition introduces a probabilistic inference function of a PKB.

Definition 4.1 *[51] (A probabilistic inference function) Let $\mathcal{R} = \langle \mathcal{B}, \mathsf{E} \rangle$ be a PKB profile, $\mathcal{K} \in \mathcal{B}$ and $\mathcal{K} \not\models \perp$. Function $\varphi_\mathcal{K}^\mathsf{E} : \mathbb{P}^{\hbar_\mathsf{E}} \to \mathbb{K}$ is called the probability rule-based deductive function of $\mathcal{K}$ over E if there exists $\vec{p} \in \mathbb{P}^{\hbar_\mathsf{E}}$ such that $\varphi_\mathcal{K}^\mathsf{E}(\vec{p}) = \mathcal{K}$.*

Intuitively, Definition 4.1 says that the probability rule-based deductive function of a PKB takes a vector of probability functions and maps them to a PKB consisting of probabilistic constraints.

Consider $\mathcal{K} \in \mathcal{B}$ over E, the maximum entropy principle-based inference means that a probability function $\mathcal{P}$ with the maximum entropy is selected among $\mathcal{P} \models \mathcal{K}$.

For any $\mathcal{P} \in \widehat{\mathcal{P}}(\mathsf{E})$, entropy of $\mathcal{P}$ is $H(\mathcal{P}) = -\sum_{\Theta \in \Lambda(\mathsf{E})} \mathcal{P}(\Theta) \cdot \log \mathcal{P}(\Theta)$.

A satisfying restored probability vector $\mathcal{K}$ over E is defined as follows:

$$\vec{\sigma}_\mathcal{K}^\mathsf{E} = \arg \max_{\mathcal{P} \in \mathcal{U}(\mathcal{K})} H(\mathcal{P}) = \arg \min_{\mathcal{P} \in \mathcal{U}(\mathcal{K})} \sum_{\Theta \in \Lambda(\mathsf{E})} \mathcal{P}(\Theta) \cdot \log \mathcal{P}(\Theta). \tag{4.1}$$

Definition 4.2 *[51] Let $\mathcal{R} = \langle \mathcal{B}, \mathsf{E} \rangle$ be a PKB profile, $\mathcal{K} \in \mathcal{B}$. A probability vector of $\mathcal{K}$ over E is $\vec{\sigma}_\mathcal{K}^\mathsf{E} = (\mathcal{P}_\mathcal{K}(\Theta_1), \dots, \mathcal{P}_\mathcal{K}(\Theta_{\hbar_\mathsf{E}}))$ iff $\sum_{i=1}^{\hbar_\mathsf{E}} \mathcal{P}_\mathcal{K}(\Theta_i) = 1$*

Definition 4.3 *[51]* *Let* $\mathcal{R} = \langle \mathcal{B}, \mathbf{E} \rangle$ *be a PKB profile,* $\mathcal{K} = \{(F_1|G_1)[\rho_1], \dots, (F_{\bar{b}_{\mathcal{K}}}|G_{\bar{b}_{\mathcal{K}}})[\rho_{\bar{b}_{\mathcal{K}}}]\} \in \mathcal{B}$ *and* $\mathcal{K} \models \perp$. *A satisfying restored probability vector of* $\mathcal{K}$ *over* $\mathbf{E}$ *is* $\vec{\sigma}_{\mathcal{K}}^{\mathbf{E}} = (\mathcal{P}_{\mathcal{K}}(\Theta_1), \dots, \mathcal{P}_{\mathcal{K}}(\Theta_{\hbar_E}))$ *iff the following conditions are satisfied:*

(i) $\sum_{i=1}^{\hbar_E} \mathcal{P}_{\mathcal{K}}(\Theta_i) = 1.$

(ii) $\exists \mathcal{K}^* = \varphi_{\mathcal{K}^*}(\vec{\sigma}_{\mathcal{K}}^{\mathbf{E}}) = \{(F_1|G_1)[\rho_1^*], \dots, \left(F_{\bar{b}_{\mathcal{K}}} \middle| G_{\bar{b}_{\mathcal{K}}}\right)[\rho_{\bar{b}_{\mathcal{K}}}^*]\}$ *such that* $\mathcal{K}^* \not\models \perp.$

Intuitively, Definition 4.3 shows that a consistent PKB can be found from a satisfying restored probability vector by using the probability rules in Theorem 2.1.

Example 4.1 *Consider* $\mathcal{K}_1$ *in example 2.3. Consider* $\mathcal{P}_{\mathcal{K}_1}(HTD) = 0.14,$ $\mathcal{P}_{\mathcal{K}_1}(HT\bar{D}) = 0.21,$ $\mathcal{P}_{\mathcal{K}_1}(H\bar{T}D) = 0.15,$ $\mathcal{P}_{\mathcal{K}_1}(H\bar{T}\bar{D}) = 0.20,$ $\mathcal{P}_{\mathcal{K}_1}(\bar{H}TD) = 0.05,$ $\mathcal{P}_{\mathcal{K}_1}(\bar{H}T\bar{D}) = 0.05,$ $\mathcal{P}(\bar{H}_{\mathcal{K}_1}\bar{T}D) = 0.11,$ $\mathcal{P}_{\mathcal{K}_1}(\bar{H}\bar{T}\bar{D}) = 0.09.$ *We have,*

$\vec{\sigma}_{\mathcal{K}_1}^{\mathbf{E}} = (0.14, 0.21, 0.15, 0.20, 0.05, 0.05, 0.11, 0.09).$

Then, $\mathcal{P}_{\mathcal{K}_1}(HTD) + \mathcal{P}_{\mathcal{K}_1}(HT\bar{D}) + \mathcal{P}_{\mathcal{K}_1}(H\bar{T}D) + \mathcal{P}_{\mathcal{K}_1}(H\bar{T}\bar{D}) + \mathcal{P}_{\mathcal{K}_1}(\bar{H}TD) + \mathcal{P}_{\mathcal{K}_1}(\bar{H}T\bar{D}) + \mathcal{P}_{\mathcal{K}_1}(\bar{H}\bar{T}D) + \mathcal{P}_{\mathcal{K}_1}(\bar{H}\bar{T}\bar{D}) = 1.$

By rule (P0) in Theorem 2.1,

$\mathcal{P}(H) = \mathcal{P}_{\mathcal{K}_1}(HTD) + \mathcal{P}_{\mathcal{K}_1}(HT\bar{D}) + \mathcal{P}_{\mathcal{K}_1}(H\bar{T}D) + \mathcal{P}_{\mathcal{K}_1}(H\bar{T}\bar{D}) = 0.7,$

$\mathcal{P}(T) = 0.43, \mathcal{P}(D) = 0.45$ *and*

$\mathcal{P}(TH) = \mathcal{P}_{\mathcal{K}_1}(HTD) + \mathcal{P}_{\mathcal{K}_1}(HT\bar{D}) = 0.35.$

By rule (P1) in Theorem 2.1, $\mathcal{P}(TH) = \mathcal{P}_{\mathcal{K}_1}(HTD) + \mathcal{P}_{\mathcal{K}_1}(HT\bar{D}) = 0.35,$ $\mathcal{P}(HD) = 0.29.$

By rule (P2) in Theorem 2.1,

$\mathcal{P}(T|H) = \frac{0.35}{0.7} = 0.5,$

$\mathcal{P}(H|D) = 0.64.$

Therefore, $\mathcal{K}_1^* = \{(H)[0.7], (T)[0.43], (D)[0.45], (T|H)[0.5], (H|D)[0.64]\}$ *and* $\mathcal{K}_1^* \not\models \perp$ *is inferred from* $\vec{\sigma}_{\mathcal{K}_1}^{\mathbf{E}}.$

4.2.2 Desired Properties of Consistency-Restoring Operator

The consistency-restoring operator is a function that maps from an inconsistent PKB to a consistent PKB. This operator should satisfy desired properties to characterize it. In this section, the desired properties are used to handle and characterize inconsistencies that appear in a PKB. The following properties are inherited from [57, 76] and modified to be suitable for the probability environment.

Definition 4.4 *([54, 57, 76])* *Let* $\mathcal{K}, \mathcal{K}_1, \mathcal{K}_2 \in \mathbb{K}$ *be PKBs. Function* $\eta : \mathbb{K} \to \mathbb{K}$ *is called a consistency-restoring operator iff the following properties are satisfied:*

(SUC)-**Success** $\forall \mathcal{K} \in \mathbb{K} : \eta(\mathcal{K}) \not\models \perp$

This property assures that all PKBs become consistent ones after the consistency-restoring process.

(SPR)-**Structure Presevation** $\forall \mathcal{K} \in \mathbb{K}$ *and there exists* $\vec{\vartheta} \in \mathbb{R}_{[0,1]}^{\bar{b}_{\mathcal{K}}}$ *such that* $\eta(\mathcal{K}) = \partial_{\mathcal{K}}(\vec{\vartheta})$

It states that the structure of constraints in a PKB when this operator is applied.

(VAC)-**Vacuity** *If* $\mathcal{K} \not\models \perp$ *then* $\eta(\mathcal{K}) = \mathcal{K}$

The property **VAC** *states that a consistent knowledge base has no modification after the consistency-restoring process.*

(IRS)-**Irrelevance of Syntax** *If* $\mathcal{K}_1 \overset{\triangle}{=} \mathcal{K}_2$ *then* $\eta(\mathcal{K}_1) \overset{\triangle}{=} \eta(\mathcal{K}_2)$

It requires that if two PKBs are semi-extensionally equivalent then they will be semi-extensionally equivalent after the consistency-restoring process.

(NOD)-**Non-Dictatorship** *If* $\kappa = (F|G)[\rho]$ *and* $G \not\equiv \top$ *then* $\exists \mathcal{K} : \kappa \in \mathcal{K}, \kappa \notin \eta(\mathcal{K})$

The above properties ensure that no probability constraint could be present in every consistency-restoring result of a PKB.

(WIA)-**Weak Irrelevant Alternatives** *If* $\textbf{SC}(\mathcal{K}_1) \cap \textbf{SC}(\mathcal{K}_2) = \emptyset$ *then* $\eta(\mathcal{K}_1 \cup \mathcal{K}_2) \overset{\triangle}{=} \eta(\mathcal{K}_1) \cup \eta(\mathcal{K}_2)$

It requires that if there are no common probabilistic constraints between two PKBs then the restoring result of their union and that of two disjoint PKBs are extensionally equivalent.

(IA)-**Irrelevant Alternatives**

If $(\eta(\mathcal{K}_1) \cup \eta(\mathcal{K}_2)) \not\models \perp$ *then* $\eta(\mathcal{K}_1) \cup \eta(\mathcal{K}_2) \overset{\triangle}{=} \eta(\mathcal{K}_1 \cup \mathcal{K}_2)$

This property IA implies that if the union of the consistency-restoring result of two disjoint PKBs is consistent then the consistency-restoring result of their union is extensionally equivalent to the union of the consistency-restoring result of two disjoint PKBs.

Theorem 4.1 ([54, 57, 76]) *If η fulfills* **SUC** *and fulfills* **IA** *then η also fulfills* **WIA**.

Proof 4.1 *Let* $\mathcal{K}_1, \mathcal{K}_2 \in \mathbb{K}$. *Since η fulfills* **SUC***,* $\eta(\mathcal{K}_1) \not\models \perp$ *and* $\eta(\mathcal{K}_2) \not\models \perp$.
Since, $\textbf{SC}(\mathcal{K}_1) \cap \textbf{SC}(\mathcal{K}_2) = \emptyset$ *and η fulfills* **SUC***,* $\eta(\mathcal{K}_1) \not\models \perp$ *and* $\eta(\mathcal{K}_2) \not\models \perp$.
Therefore, $\eta(\mathcal{K}_1) \cup \eta(\mathcal{K}_1) \not\models \perp$. *Since η fulfills* **IA***,* $\eta(\mathcal{K}_1 \cup \mathcal{K}_2) \overset{\triangle}{=} \eta(\mathcal{K}_1) \cup \eta(\mathcal{K}_2)$. $\quad\square$

4.2.3 A General Model for Restoring Consistency

The consistency-restoring problem is defined as follows:

(1) **Input:** $\mathcal{K} = \{(c_1)[\rho_1], \ldots, (c_h)[\rho_h]\}$ and E.

(2) **Output:** $\mathcal{K}^* = \{(c_1)[\vartheta_1], \ldots, (c_h)[\vartheta_h]\}$ such that $\mathcal{K}^* \not\models \perp$.

(3) **Scope of problem:** A PKB is represented by probabilistic constraints.

(4) **The consistency-restoring process:**

- **Step 1**: Computing an inconsistency measure. If it equals zero then stop. Otherwise, go to step 2.

- **Step 2**: Solving inconsistencies.

- **Step 3**: Calculating new probability for each probabilistic constraint by using the consistency-restoring operator.

A general model for restoring the consistency of a PKB is presented in Figure 4.1.

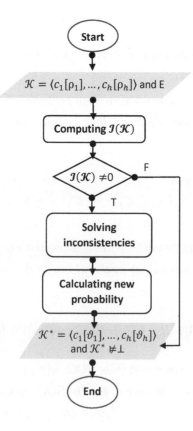

Figure 4.1 A general model for restoring consistency.

4.3 METHODS FOR RESTORING CONSISTENCY

4.3.1 The Norm-based Consistency-restoring Problem

A model for restoring the norm-based consistency of a PKB is presented in Figure 4.2. It consists of three steps:

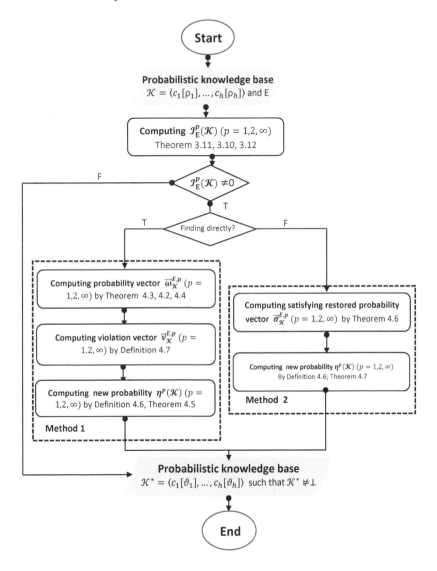

Figure 4.2 A model for restoring the norm-based consistency.

Step 1: Computing the inconsistency measure with respect to 1-norm by using Theorem 3.11, 2-norm by using Theorem 3.10, ∞-norm by using Theorem 3.12. If it equals zero then stop. Otherwise, go to step 2.

Step 2: Solving inconsistencies.

- Method 1:

+ Finding probability vector of $\mathcal{K}$: with respect to 1-norm by using Theorem 4.3, 2-norm by using Theorem 4.2 and ∞-norm by Theorem 4.4.

+ Finding the violation vector by using Definition 4.6.

- Method 2: Finding satisfying restored probability vector of $\mathcal{K}$ by using Theorem 4.6.

Step 3: Calculating new probability for each probabilistic constraint

- Method 1: Using Definition 4.5 and Theorem 4.5.

- Method 2: Using Definition 4.5 and Theorem 4.7.

From the model, it is easy to see that both methods look for new probabilities of probabilistic constraints such that the resulting PKB is consistent. However, with the first method, the new probability of probabilistic constraints could be directly found by solving the unconstrained optimization problem. This problem is built from the violation vector and the probability vector of a PKB. For with the second method, the new probability of probabilistic constraints could be found by applying probability rules on the satisfying restored probability vector.

First of all, Definition 4.5 represents the consistency-restoring operator with respect to p-norm in order to evaluate this model.

Definition 4.5 ([51, 54]) *Let $\mathcal{R} = \langle \mathcal{B}, \mathsf{E} \rangle$ be a PKB profile, $\mathcal{K} \in \mathcal{B}$, and $\bar{\mathcal{P}} \in \vec{\sigma}_{\mathcal{K}}^{\mathsf{E}}$. The consistency-restoring operator $\eta^p : \mathbb{K} \to \mathbb{K}$ with respect to p-norm $(p \geq 1)$ of $\mathcal{K}$ is defined as follows:*

$$\eta^p(\mathcal{K}) = \partial_{\mathcal{K}}(\vec{\vartheta}) \qquad (4.2)$$

where, $\vec{\vartheta} = (\vartheta_1, \ldots, \vartheta_{\bar{b}_{\mathcal{K}}})$ with

$$\vartheta_i = \begin{cases} \bar{\mathcal{P}}(F_i \,|\, G_i), & \text{if } \bar{\mathcal{P}}(G_i) > 0. \\ \rho_i, & \text{otherwise.} \end{cases} \qquad (4.3)$$

1) Method 1

In this subsection, we will present several theorems as the basis for building the consistency-restoring model according to the first method in Figure 4.2. Theorem 4.2 is employed to find the probability vector with respect to p-norm $(p > 1$ and $p \neq \infty)$ of a PKB.

Theorem 4.2 ([51, 54]) *Let $\mathcal{R} = \langle \mathcal{B}, \mathsf{E} \rangle$ be a PKB profile and $\mathcal{K} \in \mathcal{B}$.*

Let $f : \mathbb{R}^{\hbar_E} \to \mathbb{R}^*$ such that $f(\vec{\omega}) = \left\| A_{\mathcal{K}}^{E} \vec{\omega} \right\|_p$. For the following optimization problem:

$$\arg \min_{\vec{\omega} \in \mathbb{R}^{\hbar_E}} \left\| A_{\mathcal{K}}^{E} \vec{\omega} \right\|_p \tag{4.4}$$

subject to $\vec{\omega} \in C_p$.

There always exists an optimal solution $\vec{\omega}^*$; simultaneously $\vec{\omega}^*$ is a probability vector of $\mathcal{K} \in \mathcal{B}$ over E.

Proof 4.2 As in the proof of Theorem 3.10, Problem (4.4) is always feasible. This indicates that $\exists \vec{\omega}^* \in \mathbb{R}^{\hbar_E}$ such that $\sum_{j=1}^{\hbar_E} \omega_j^* = 1, \omega_j^* \geq 0 \ \forall j = \overline{1, \hbar_E}$. Therefore, by Definition 4.2, $\vec{\omega}^*$ is a probability vector of $\mathcal{K} \in \mathcal{B}$ over E. $\qquad\square$

We call the solution of Problem (4.4) a *probability vector with respect to p-norm* ($p > 1$ and $p \neq \infty$) of $\mathcal{K}$ over E, denoted by $\vec{\omega}_{\mathcal{K}}^{E,p}$.

Theorem 4.3 is employed to find a probability vector with respect to 1-norm of a PKB.

Theorem 4.3 ([51, 54]) *Let $\mathcal{R} = \langle \mathcal{B}, \mathsf{E} \rangle$ be a PKB profile and $\mathcal{K} \in \mathcal{B}$. Let $f : \mathbb{R}^{\hbar_E + \bar{b}_{\mathcal{K}}} \to \mathbb{R}^*$ such that $f(\vec{\omega}, \vec{\lambda}) = \sum_{i=1}^{\bar{b}_{\mathcal{K}}} \lambda_i$. For the following linear optimization problem:*

$$\arg \min_{(\vec{\omega}, \vec{\lambda}) \in \mathbb{R}^{\hbar_E + \bar{b}_{\mathcal{K}}}} \sum_{i=1}^{\bar{b}_{\mathcal{K}}} \lambda_i \tag{4.5}$$

subject to $(\vec{\omega}, \vec{\lambda}) \in C_1$

There always exists an optimal solution $\vec{\omega}^*$, simultaneously $\vec{\omega}^*$ is a probability vector of $\mathcal{K} \in \mathcal{B}$ over E.

Proof 4.3 The proof is similar to the proof of Theorem 4.2.

Problem (4.5) is always feasible so $\vec{\omega}^* = (\mathcal{P}(\Theta_1), \ldots, \mathcal{P}(\Theta_{\hbar_E}))$. Therefore, by Definition 4.2, $\vec{\omega}^*$ is a probability vector of $\mathcal{K} \in \mathcal{B}$ over E. $\qquad\square$

We call the solution of Problem (4.5) a *probability vector with respect to 1-norm* of $\mathcal{K}$ over E, denoted by $\vec{\omega}_{\mathcal{K}}^{E,1}$.

Theorem 4.4 is employed to find a probability vector with respect to ∞-norm of a PKB.

Theorem 4.4 ([51, 54]) *Let $\mathcal{R} = \langle \mathcal{B}, \mathsf{E} \rangle$ be a PKB profile and $\mathcal{K} \in \mathcal{B}$. Let $f : \mathbb{R}^{\hbar_E + 1} \to \mathbb{R}^*$ such that $f(\vec{\omega}, \lambda) = \lambda$. For the following linear optimization problem:*

$$\arg \min_{(\vec{\omega}, \lambda) \in \mathbb{R}^{\hbar_E + 1}} \lambda \tag{4.6}$$

subject to $(\vec{\omega}, \lambda) \in C_\infty$.

There always exists an optimal solution $\vec{\omega}^*$, simultaneously $\vec{\omega}^*$ is a probability vector of $\mathcal{K} \in \mathcal{B}$ over E.

Proof 4.4 *The proof is similar to the proof of Theorem 4.2. Problem (4.6) is always feasible so $\vec{\omega}^*=(\mathcal{P}(\Theta_1),\ldots,\mathcal{P}(\Theta_{\hbar_\mathsf{E}}))$. Therefore, by Definition 4.2, $\vec{\omega}$ is a probability vector of $\mathcal{K} \in \mathcal{B}$ over E.* □

We call the solution of Problem (4.5) *a probability vector with respect to ∞-norm* of $\mathcal{K}$ over E, denoted by $\vec{\omega}_{\mathcal{K}}^{\mathsf{E},\infty}$.

Definition 4.6 is employed to represent a violation vector with respect to p-norm $(p \geq 1)$ of a PKB.

Definition 4.6 ([54]) *Let $\mathcal{R} = \langle \mathcal{B}, \mathsf{E} \rangle$ be a PKB profile and $\mathcal{K} \in \mathcal{B}$. A violation vector $\vec{v}_{\mathcal{K}}^{\mathsf{E},p}$ with respect to p-norm $(p \geq 1)$ of $\mathcal{K}$ over E is defined as follows:*

$$\vec{v}_{\mathcal{K}}^{\mathsf{E},p} = A_{\mathcal{K}}^{\mathsf{E}} \cdot \vec{\omega}_{\mathcal{K}}^{\mathsf{E},p} \tag{4.7}$$

Example 4.2 *Let's continue example 3.12. By Definition 4.6, we have:*

$$\vec{v}_{\mathcal{K}_1}^{\mathsf{E},1} = \begin{pmatrix} 0.3 & 0.3 & 0.3 & 0.3 & -0.7 & -0.7 & -0.7 & -0.7 \\ 0.7 & 0.7 & -0.3 & -0.3 & 0.7 & 0.7 & -0.3 & -0.3 \\ 0.55 & -0.45 & 0.55 & -0.45 & 0.55 & -0.45 & 0.55 & -0.45 \\ 0.5 & -0.5 & 0.5 & -0.5 & 0 & 0 & 0 & 0 \\ 0.36 & 0 & 0.36 & 0 & -0.64 & 0 & -0.64 & 0 \end{pmatrix} \begin{pmatrix} 0 \\ 0.35 \\ 0.29 \\ 0.06 \\ 0.1 \\ 0 \\ 0.06 \\ 0.14 \end{pmatrix}$$

Then, $\vec{v}_{\mathcal{K}_1}^{\mathsf{E},1} = (0, 0.05, 0, 0, 0)$. As $\mathcal{K}_2$, $\mathcal{K}_3$, $\mathcal{K}_4$ is consistent so violation vectors of these PKBs is $\vec{0}$. Violation vectors of $\mathcal{K}_1$, $\mathcal{K}_2$, $\mathcal{K}_3$, $\mathcal{K}_4$, $\mathcal{K}_5$ with respect to 1-norm, 2-norm and ∞-norm are presented in Table 4.2.

Table 4.2 Violation vectors $\mathcal{K}_1, \mathcal{K}_2, \mathcal{K}_3, \mathcal{K}_4, \mathcal{K}_5$ with respect to p-norm

$\vec{v}_{\mathcal{K}_1}^{\mathsf{E},1}$	$\vec{v}_{\mathcal{K}_2}^{\mathsf{E},1}$	$\vec{v}_{\mathcal{K}_3}^{\mathsf{E},1}$	$\vec{v}_{\mathcal{K}_4}^{\mathsf{E},1}$	$\vec{v}_{\mathcal{K}_5}^{\mathsf{E},1}$	$\vec{v}_{\mathcal{K}_1}^{\mathsf{E},2}$	$\vec{v}_{\mathcal{K}_2}^{\mathsf{E},2}$	$\vec{v}_{\mathcal{K}_3}^{\mathsf{E},2}$	$\vec{v}_{\mathcal{K}_4}^{\mathsf{E},2}$	$\vec{v}_{\mathcal{K}_5}^{\mathsf{E},2}$	$\vec{v}_{\mathcal{K}_1}^{\mathsf{E},\infty}$	$\vec{v}_{\mathcal{K}_2}^{\mathsf{E},\infty}$	$\vec{v}_{\mathcal{K}_3}^{\mathsf{E},\infty}$	$\vec{v}_{\mathcal{K}_4}^{\mathsf{E},\infty}$	$\vec{v}_{\mathcal{K}_5}^{\mathsf{E},\infty}$
0	0	0	0	0	-0.01	0	0	0	-0.01	-0.02	0	0	0	-0.01
0.05	0	0	0	0.03	0.02	0	0	0	0.01	0.02	0	0	0	0.01
0	0	0	0	0	0	0	0	0	0	0.02	0	0	0	0.01
0	0	0	0	0	-0.02	0	0	0	-0.01	-0.02	0	0	0	-0.01
0	0	-	-	0	0	-	-	0	0	0.02	0	-	-	0.01

Definition 4.7 represents an exponential constraint vector of a PKB.

Definition 4.7 ([54]) *Let $\mathcal{R} = \langle \mathcal{B}, \mathsf{E} \rangle$ be a PKB profile. An exponential constraint vector $\vec{\alpha}_{\mathcal{K}}^{\mathsf{E}} : \mathbb{R}^{\bar{b}_{\mathcal{K}}+1} \to \mathbb{R}^{\hbar_{\mathsf{E}}}$ of $\mathcal{K} \in \mathcal{B}$ over E is defined as follows:*

$$\vec{\alpha}_{\mathcal{K}}^{\mathsf{E}}(\vec{x}, y) = (\alpha_1(\vec{x}, y), \dots, \alpha_{\hbar_{\mathsf{E}}}(\vec{x}, y))^T \tag{4.8}$$

where

$$\alpha_j(\vec{x}, y) = \exp\left(\sum_{i=1}^{\bar{b}_{\mathcal{K}}} x_i b_{ij} + y - 1\right) \quad \forall j = \overline{1, \hbar_{\mathsf{E}}}$$

Theorem 4.5 is employed to find the consistency-restoring operator η^p with respect to p-norm $(p \geq 1)$.

Theorem 4.5 ([54]) *Let $\mathcal{R} = \langle \mathcal{B}, \mathsf{E} \rangle$ be a PKB profile, $\mathcal{K} \in \mathcal{B}$. There exists $\vec{\vartheta} = (\vartheta_1, \dots, \vartheta_{\bar{b}_{\mathcal{K}}})$ that such $\eta^p(\mathcal{K}) = \partial_{\mathcal{K}}(\vec{\vartheta})$, where $\vec{\vartheta}$ corresponds to the solution $\vec{x}^{p*}$ of the following unconstrained optimization problem:*

$$\arg\min_{(\vec{x}, y) \in \mathbb{R}^{\bar{b}_{\mathcal{K}}+1}} \sum_{j=1}^{\hbar_{\mathsf{E}}} \left(\vec{\alpha}_{\mathcal{K}}^{\mathsf{E}}(\vec{x}, y)\right)_j - \vec{x}^\top \vec{v}_{\mathcal{K}}^{\mathsf{E},p} - y \tag{4.9}$$

Proof 4.5 *We will prove that (a) Problem (4.9) has a feasible solution and (b) $\vec{x}^{p*}$ is a vector such that $\eta^p(\mathcal{K}) = \partial_{\mathcal{K}}(\vec{x}^{p*})$.*

(a) Firstly, it is easy to see that Problem (4.9) is always feasible. We have a constraint empty set. The objective function $f(\vec{x}, y) = \sum_{j=1}^{\hbar_{\mathsf{E}}} \left(\vec{\alpha}_{\mathcal{K}}^{\mathsf{E}}(\vec{x}, y)\right)_j - \vec{x}^\top \vec{v}_{\mathcal{K}}^p - y$ is the exponential function so $f(\vec{x}, y)$ is a convex function [68, 56]. Therefore, Problem (4.9) is the convex optimization problem. Assume that $\mathcal{G}$ is the optimal solution of Problem (4.9), we have $\vec{x}^{p*} \in (\vec{x}^{p*}, y^*)$ with $\mathcal{G} = \sum_{j=1}^{\hbar_{\mathsf{E}}} \left(\vec{\alpha}_{\mathcal{K}}^{\mathsf{E}}(\vec{x}^*, y^*)\right)_j - \vec{x}^{*\top} \vec{v}_{\mathcal{K}}^p - y^*$ and $(\vec{x}^{p*}, y^{p*})$ is an optimal point.

As the minimum of convex function in a convex set is unique so $(\vec{x}^{p*}, y^{p*})$ is uniquely feasible and attains the optimal value g^{p*} [68]. Therefore, $(\vec{x}^{p*}, y^{p*})$ is an optimal solution of Problem (4.14) [68].

(b) Secondly, if $\mathcal{K}$ is consistent then $\vec{v}_{\mathcal{K}}^{\mathsf{E},p} = 0$. By Definition 4.6 $\vec{x}^{p*} = \vec{\omega}_{\mathcal{K}}^{\mathsf{E},p}$. Hence, $\eta^p(\mathcal{K}) = \partial_{\mathcal{K}}(\vec{x}^{p*})$

Conversely, if $\mathcal{K}$ is inconsistent then $\vec{v}_{\mathcal{K}}^{\mathsf{E},p} \neq 0$. Problem (4.9) offers an optimal solution $(\vec{x}^{p*}, y^{p*})$ such that $\vec{x}^{*\top} \vec{v}_{\mathcal{K}}^p - y^*$ makes $\mathcal{K}$ consistent. Then, there exists $\mathcal{K}^* = \varphi_{\mathcal{K}^*}(\vec{x}^{p*})$ such that $\mathcal{K}^* \not\models \bot$. □

Example 4.3 *Let's continue example 3.2 and Example 4.2, by Table 4.2 and by Theorem 4.5*

- *By Definition 4.7, we have* $\vec{\alpha}^E_{\mathcal{K}_1}(\vec{x}, y) = (\alpha_1(\vec{x}, y), \ldots, \alpha_8(\vec{x}, y))^T$, *where*

$\alpha_1(\vec{x}, y) = \exp\left(0.3x_1 + 0.7x_2 + 0.55x_3 + 0.5x_4 + 0.36x_5 + y - 1\right),$

$\alpha_2(\vec{x}, y) = \exp\left(0.3x_1 + 0.7x_2 - 0.45x_3 - 0.5x_4 + y - 1\right),$

$\alpha_3(\vec{x}, y) = \exp\left(0.3x_1 - 0.3x_2 + 0.55x_3 + 0.5x_4 + 0.36x_5 + y - 1\right),$

$\alpha_4(\vec{x}, y) = \exp\left(0.3x_1 - 0.3x_2 - 0.45x_3 - 0.5x_4 + y - 1\right),$

$\alpha_5(\vec{x}, y) = \exp\left(-0.7x_1 + 0.7x_2 + 0.55x_3 - 0.64x_5 + y - 1\right),$

$\alpha_6(\vec{x}, y) = \exp\left(-0.7x_1 + 0.7x_2 - 0.45x_3 + y - 1\right),$

$\alpha_7(\vec{x}, y) = \exp\left(-0.7x_1 - 0.3x_2 + 0.55x_3 - 0.64x_5 + y - 1\right),$

$\alpha_8(\vec{x}, y) = \exp\left(-0.7x_1 - 0.3x_2 - 0.45x_3 + y - 1\right)$

- *For 1-norm,* $\vec{\vartheta}$ *corresponds to* $\vec{x}^{1*}$ *is computed by solving the unconstrained optimization problem:*

$$\arg\min_{(\vec{x},y)\in\mathbb{R}^{5+1}} \sum_{i=1}^{8} \alpha_i(\vec{x}, y) + 0.5x_2 - y \tag{4.10}$$

- *For 2-norm,* $\vec{\vartheta}$ *corresponds to* $\vec{x}^{2*}$ *is computed by solving the unconstrained optimization problem:*

$$\arg\min_{(\vec{x},y)\in\mathbb{R}^{5+1}} \sum_{i=1}^{8} \alpha_i(\vec{x}, y) - 0.01x_1 + 0.02x_2 - 0.02x_4 - y \tag{4.11}$$

- *For ∞-norm,* $\vec{\vartheta}$ *corresponds to* $\vec{x}^{\infty*}$ *is computed by solving the unconstrained optimization problem:*

$$\arg\min_{(\vec{x},y)\in\mathbb{R}^{5+1}} \sum_{i=1}^{8} \alpha_i(\vec{x}, y) + 0.02(-x_1 + x_2 + x_3 - x_4 + x_5) - y \tag{4.12}$$

Solving problem (4.10), we have $\vec{\vartheta} = \vec{x}^* = (0.7, 0.36, 0.45, 0.5, 0.64)$. *Therefore,* $\mathcal{K}^* = \eta^1(\mathcal{K}_1) = \{(H)[0.7], (T)[0.36], (D)[0.45], (T|H)[0.5], (H|D)[0.64]\}$ *is consistent.*

After using consistency-restoring operators $\eta^1, \eta^2, \eta^\infty$, *new probability values of constraints in* $\mathcal{K}_1$ *and* $\mathcal{K}_5$ *with respect to 1-norm, 2-norm, ∞-norm are shown in Table 4.3. For 1-norm, the probability value of $(T)[0.3]$ in $\mathcal{K}_1$ and that of $(T)[0.31]$ in $\mathcal{K}_5$ have been changed. For 2-norm, the probability value of $(H)[0.7]$, $(T)[0.3]$, $(T|H)[0.5]$ in $\mathcal{K}_1$ and that of $(H)[0.8]$, $(T)[0.31]$, $(T|H)[0.42]$ in $\mathcal{K}_5$ have been changed. For ∞-norm, the probability value of all probabilistic constraints in $\mathcal{K}_1$ and $\mathcal{K}_5$ have been changed. This change makes $\mathcal{K}_1$ and $\mathcal{K}_5$ consistent.*

Table 4.3 The probability value of all probabilistic constraints in $\mathcal{K}_1$ and $\mathcal{K}_5$ based on an unconstrained optimization problem

$\kappa_i \in \mathcal{K}_1$	$\eta^1(\mathcal{K}_1)$	$\eta^2(\mathcal{K}_1)$	$\eta^\infty(\mathcal{K}_1)$	$\kappa_i \in \mathcal{K}_5$	$\eta^1(\mathcal{K}_5)$	$\eta^2(\mathcal{K}_5)$	$\eta^\infty(\mathcal{K}_5)$
$(H)[0.7]$	0.7	0.68	0.69	$(H)[0.8]$	0.8	0.78	0.69
$(T)[0.3]$	0.36	0.33	0.31	$(T)[0.31]$	0.35	0.36	0.34
$(D)[0.45]$	0.45	0.45	0.46	$(D)[0.5]$	0.5	0.5	0.51
$(T\,|\,H)[0.5]$	0.5	0.48	0.48	$(T\,|\,H)[0.42]$	0.42	0.42	0.4
$(H\,|\,D)[0.64]$	0.64	0.64	0.65	$(H\,|\,D)[0.6]$	0.6	0.6	0.67

2) Method 2

In this subsection, we will present several theorems as the basis for building the consistency-restoring model according to the second method in Figure 4.2.

We make Definition 4.8 to represent a symmetric probability vector of a PKB.

Definition 4.8 *Let* $\mathcal{R} = \langle \mathcal{B}, \mathbf{E} \rangle$ *be a PKB profile,* $\mathcal{K} \in \mathcal{B}$ *and* $\vec{x} \in \mathbb{R}_{[0,1]}^{\hbar_E}$. *A symmetric probability vector of* $\mathcal{K}$ *is* $\vec{z}_{\mathcal{K}}^{\mathbf{E}} = (z_1, \ldots, z_{\bar{b}_{\mathcal{K}}})^T$, *where* $\forall i = \overline{1, \bar{b}_{\mathcal{K}}}$, z_i *is defined as follows:*

$$z_i = \begin{cases} \rho_i & \text{if } \kappa_i \in \mathcal{K} \text{ and } \kappa_i = (F_i)\,[\rho_i]. \\ \rho_i \sum_{\Theta_j | \Theta_j \models G_i} x_j & \text{if } \kappa_i \in \mathcal{K} \text{ and } \kappa_i = (F_i | G_i)\,[\rho_i]. \end{cases} \quad (4.13)$$

Theorem 4.6 is employed to find the satisfying restored probability vector with respect to p-norm ($p > 1$ and $p \neq \infty$) of a PKB.

Theorem 4.6 ([51]) *Let* $\mathcal{R} = \langle \mathcal{B}, \mathbf{E} \rangle$ *be a PKB profile,* $\mathcal{K} \in \mathcal{B}$, ϵ^p *be an inconsistency measure of* $\mathcal{K}$ *with respect to p-norm (p ≥ 1) and* $\vec{z}_{\mathcal{K}}^{\mathbf{E}} = (z_1, \ldots, z_{\bar{b}_{\mathcal{K}}})^\top$ *be a symmetric probability vector of* $\mathcal{K}$. *Let* $g : \mathbb{R}^{\hbar_E + \bar{b}_{\mathcal{K}}} \to \mathbb{R}^*$ *such that* $g(\vec{x}, \vec{y}) = \sum_{j=1}^{\hbar_E} x_j \cdot log(x_j)$. *A satisfying restored probability vector* $\vec{\sigma}_{\mathcal{K}}^{E,p}$ *with respect to p-norm of* $\mathcal{K}$ *corresponds to* $\vec{x}^{p*}$ *that is a part of solution of the non-linear optimization problem:*

$$\arg \min_{(\vec{x}, \vec{y}) \in \mathbb{R}^{\hbar_E + \bar{b}_{\mathcal{K}}}} \sum_{j=1}^{\hbar_E} x_j \cdot log(x_j) \quad (4.14)$$

subject to $(\vec{x}, \vec{y}) \in C_r$

where, $C_r = \left\{ (\vec{x}, \vec{y}) \in \mathbb{R}^{\hbar_E + \bar{b}_{\mathcal{K}}} | C_{\mathcal{K}}^{E,+} \cdot \vec{x} = \vec{y} + \vec{z}_{\mathcal{K}}^{\mathbf{E}}, \sum_{j=1}^{\hbar_E} x_j = 1, \sum_{i=1}^{\bar{b}_{\mathcal{K}}} y_i = \epsilon^p, \vec{x} \geq \vec{0}, \vec{y} \geq \vec{0} \right\}$ *is a set of constraints of the norm-based consistency-restoring problem.*

Proof 4.6 *We will prove that (a) Problem (4.14) has a feasible solution, and (b)* $\vec{x}^{p*}$ *is the satisfying restored probability vector of* $\mathcal{K}$ *over* $\mathbf{E}$.

(a) Firstly, it is easy to see that Problem (4.14) is always feasible. The set C_r consists of $\hbar_E + 2\bar{b}_K + 2$ equality constraint functions. As equality functions are convex, these functions in C_r are also convex [68, 56]. The objective function $g(\vec{x}, \vec{y}) = \sum_{j=1}^{\hbar_E} x_j \cdot log(x_j)$ is the entropy function on the set of positive numbers so $g(\vec{x}, \vec{y})$ is a convex function [68, 56]. Therefore, Problem (4.14) is the convex optimization problem. Assume that $\mathcal{G}$ is the optimal solution of Problem (4.14), we have $\vec{x}^{p*} \in (\vec{x}^{p*}, \vec{y}^{p*})$ with $\mathcal{G} = \sum_{i=1}^{\hbar_E} x_i^{p*} \cdot log(x_i^{p*})$ and $(\vec{x}^{p*}, \vec{y}^{p*})$ is an optimal point.

As the minimum of convex function in a convex set is unique, so $(\vec{x}^{p*}, \vec{y}^{p*})$ is uniquely feasible and attains the optimal value g^{p*} [68]. Therefore, $(\vec{x}^{p*}, \vec{y}^{p*})$ is the optimal solution of Problem (4.14) [68].

(b) Secondly, if $\mathcal{K}$ is consistent then Problem (4.14) has an optimal solution $(\vec{x}^{p*}, \vec{y}^{p*})$ such that $\sum_{i=1}^{\bar{b}_K} y_i^{p*} = 0$. As $\vec{x}^{p*} \in C_r$ so $\sum_{j=1}^{\hbar_E} x_j^{p*} = 1$. By Definition 4.1, we have $\varphi_K^E(\vec{x}^{p*}) = \mathcal{K}$. Therefore, by Definition 4.3, $\vec{x}^{p*}$ is the satisfying restored probability vector of $\mathcal{K}$ over E.

Conversely, if $\mathcal{K}$ is inconsistent then Problem (4.14) has an optimal solution $(\vec{x}^{p*}, \vec{y}^{p*})$, such that $\sum_{i=1}^{\bar{b}_K} y_i^{p*} = \epsilon^p$. As $\vec{x}^{p*} \in C_r$ so $\sum_{j=1}^{\hbar_E} x_j^{p*} = 1$. Simultaneously, $C_K^{E,+} \cdot \vec{x}^{p*} = \vec{y}^{p*} + \vec{z}_K^E$. Then, there exists $\mathcal{K}^* = \varphi_{\mathcal{K}^*}(\vec{x}^{p*})$ such that $\mathcal{K}^* \not\models \perp$. Therefore, by Definition 4.3, $\vec{x}^{p*}$ is the satisfying restored probability vector of $\mathcal{K}$ over E. □

Theorem 4.7 is employed to determine the consistency-restoring operator with respect to p-norm of a PKB.

Theorem 4.7 ([51]) *Let $\mathcal{R} = \langle \mathcal{B}, E \rangle$ be a PKB profile, $\mathcal{K} \in \mathcal{B}$. There exists the satisfying restored probability vector $\vec{\sigma}_K^{E,p}$ of $\mathcal{K}$ then the consistency-restoring operator η^p with respect to p-norm ($p \geq 1$) of $\mathcal{K}$ is determined as follows:*

$$\eta^p(\mathcal{K}) = \varphi_{\mathcal{K}^*}^E(\vec{\sigma}_K^{E,p}) = \mathcal{K}^* \tag{4.15}$$

Proof 4.7 *If there exists the satisfying restored probability vector $\vec{\sigma}_K^{E,p}$ of $\mathcal{K}$ then by Definition 4.3 there exists $\mathcal{K}^* = \varphi_{\mathcal{K}^*}(\vec{\sigma}_K^{E,p})$ and $\mathcal{K}^* \not\models \perp$. Then, $\exists \vec{\rho}_{\mathcal{K}^*} = (\rho_1, \ldots, \rho_{\bar{b}_{\mathcal{K}^*}})$, where ρ_i corresponds to the probability values of κ_i. By Definition 4.5, $\eta^p(\mathcal{K}) = \partial_K(\vec{\rho}_{\mathcal{K}^*})$, that is, $\eta^p(\mathcal{K}) = \varphi_{\mathcal{K}^*}(\vec{\sigma}_K^{E,p}) = \mathcal{K}^*$.* □

Example 4.4 *Let's continue example 3.12. From Table 3.2, $\epsilon^1 = \mathcal{I}_{\mathcal{K}_1}^1 = 0.05$. By Definition 4.8, a symmetric probability vector of $\mathcal{K}_1$ is*
$$\vec{z}_{\mathcal{K}_1}^E = (0.7, 0.3, 0.45, 0.5(x_1 + x_2 + x3 + x4), 0.64(x_1 + x_3 + x_5 + x_7))^\top.$$

By Theorem 4.6, $\vec{x}_{\mathcal{K}}^{1*}$ of $\mathcal{K}_1$ corresponding to $\vec{x}^{1*}$ is the solution of the optimization problem:

$$\arg \min_{(\vec{x}, \vec{y}) \in \mathbb{R}^{8+5}} \sum_{i=1}^{8} x_i \cdot log(x_i) \tag{4.16}$$

subject to:

$$x_1 + x_2 + x_3 + x_4 = y_1 + 0.7, x_1 + x_2 + x_5 + x_6 = y_2 + 0.3 \tag{4.17}$$

$$x_1 + x_3 + x_5 + x_7 = y_3 + 0.45, x_1 + x_2 = 0.5(x_1 + x_2 + x_3 + x_4) + y_4 \tag{4.18}$$

$$x_1 + x_3 = 0.64(x_1 + x_3 + x_5 + x_7) + y_5 \tag{4.19}$$

$$x_1 + x_2 + x_3 + x_4 + x_5 + x_6 + x_7 + x_8 = 1 \tag{4.20}$$

$$y_1 + y_2 + y_3 + y_4 + y_5 = 0.05 \tag{4.21}$$

$$x_1 \geq 0, x_2 \geq 0, x_3 \geq 0, x_4 \geq 0, x_5 \geq 0, x_6 \geq 0, x_7 \geq 0, x_8 \geq 0 \tag{4.22}$$

$$y_1 \geq 0, y_2 \geq 0, y_3 \geq 0, y_4 \geq 0, y_5 \geq 0 \tag{4.23}$$

The solution of Problem (4.16) is $(\vec{x}_{\mathcal{K}_1}^{1*}, \vec{y}_{\mathcal{K}_1}^{1*})$,

where

$\vec{x}_{\mathcal{K}_1}^{1*} = (0.14, 0.21, 0.15, 0.20, 0.05, 0.05, 0.11, 0.09)$ *and*

$\vec{y}_{\mathcal{K}_1}^{1*} = (0, 0.05, 0, 0, 0).$

Therefore, the satisfying restored probability vector with respect to 1-norm of $\mathcal{K}_1$ is $\vec{\sigma}_{\mathcal{K}}^{E,1} = \vec{x}_{\mathcal{K}_1}^{1*} = (0.14, 0.21, 0.15, 0.20, 0.05, 0.05, 0.11, 0.09).$

Table 4.4 shows satisfying restored probability vectors with respect to 1-norm, 2-norm and ∞-norm of $\mathcal{K}_1$ and $\mathcal{K}_5$.

Table 4.4 Satisfying restored probability vectors with respect to 1-norm, 2-norm, ∞-norm of $\mathcal{K}_1$ and $\mathcal{K}_5$ [51]

Norm	Vectors	HTD	$HT\bar{D}$	$H\bar{T}D$	$H\bar{T}\bar{D}$	$\bar{H}TD$	$\bar{H}T\bar{D}$	$\bar{H}\bar{T}D$	$\bar{H}\bar{T}\bar{D}$	Sum
1-norm	$\vec{\sigma}_{\mathcal{K}_1}^{E}$	0.14	0.21	0.15	0.20	0.05	0.05	0.11	0.09	1.0
	$\vec{\sigma}_{\mathcal{K}_5}^{E}$	0.15	0.19	0.21	0.26	0.03	0.01	0.11	0.04	1.0
2-norm	$\vec{\sigma}_{\mathcal{K}_1}^{E}$	0.14	0.21	0.14	0.21	0.04	0.04	0.12	0.10	1.0
	$\vec{\sigma}_{\mathcal{K}_5}^{E}$	0.22	0.12	0.11	0.35	0.01	0.01	0.17	0.01	1.0
∞-norm	$\vec{\sigma}_{\mathcal{K}_1}^{E}$	0.15	0.20	0.14	0.21	0.04	0.03	0.12	0.11	1.0
	$\vec{\sigma}_{\mathcal{K}_5}^{E}$	0.15	0.19	0.20	0.26	0.03	0.01	0.12	0.04	1.0

By Definition 4.3 and Theorem 2.1, for $(H)[\vartheta_1]$, we have $\vartheta_1 = P(HTD) + P(HT\bar{D}) + P(H\bar{T}D) + P(H\bar{T}\bar{D}) = 0.14 + 0.21 + 0.15 + 0.20 = 0.7$. Similarly, for $(T)[\vartheta_2]$, $\vartheta_2 = 0.14 + 0.21 + 0.05 + 0.05 = 0.45$. For $(D)[\vartheta_3]$, $\vartheta_3 = 0.14 + 0.15 + 0.05 +$

$0.11 = 0.45$. *For* $(T|H)[\vartheta_4]$, $\vartheta_4 = \frac{P(HTD)+P(HT\bar{D})}{\vartheta_1} = \frac{0.14+0.21}{0.7} = 0.5$. *Similarly, for* $(H|D)\vartheta_5$, $\vartheta_5 = 0.64$.

By Theorem 4.7,

$$\eta^1(\mathcal{K}_1) = \{(H)[0.7], (T)[0.45], (D)[0.45], (T|H)[0.5], (H|D)[0.64]\}.$$

It is easy to see that the probability value of $(T)[0.3]$ *in* $\mathcal{K}_1$ *have been changed after the consistency-restoring process. This change makes* $\mathcal{K}_1$ *consistent. New probability values of probabilistic constraints in* $\mathcal{K}_1$ *and* $\mathcal{K}_5$ *are shown in Table 4.5 after consistency-restoring operators* $\eta^1, \eta^2, \eta^\infty$ *are employed.*

Table 4.5 New probability values of probabilistic constraints in $\mathcal{K}_1$ and $\mathcal{K}_5$ with respect to the norm-based consistency-restoring operators [51]

$\kappa_i \in \mathcal{K}_1$	$\eta^1(\mathcal{K}_1)$	$\eta^2(\mathcal{K}_1)$	$\eta^\infty(\mathcal{K}_1)$	$\kappa_i \in \mathcal{K}_5$	$\eta^1(\mathcal{K}_5)$	$\eta^2(\mathcal{K}_5)$	$\eta^\infty(\mathcal{K}_5)$
$(H)[0.7]$	0.7	0.7	0.7	$(H)[0.8]$	0.8	0.8	0.8
$(T)[0.3]$	0.45	0.43	0.42	$(T)[0.31]$	0.38	0.36	0.37
$(D)[0.45]$	0.45	0.45	0.45	$(D)[0.5]$	0.5	0.51	0.5
$(T|H)[0.5]$	0.5	0.5	0.5	$(T|H)[0.42]$	0.42	0.43	0.42
$(H|D)[0.64]$	0.64	0.64	0.64	$(H|D)[0.6]$	0.71	0.65	0.69

Theorem 4.8 presents the relationship between the norm-based consistency-restoring operators and desirable properties.

Theorem 4.8 ([54]) *If* $p \geq 1$ *then the norm-based consistency-restoring operator* η^p *satisfies* **SPR, SUC, VAC, IRS, WIA, IA**. *But it only satisfies* **NOD** *when* $p > 1$.

Proof 4.8 *(SUC). By Theorem 4.7, we have* $\eta^p(\mathcal{K}) = \varphi^E_{\mathcal{K}^*}(\vec{\sigma}^{E,p}_\mathcal{K}) = \mathcal{K}^*$ *and* $\mathcal{K}^* \not\models \bot$. *Therefore,* $\eta^p(\mathcal{K}) \not\models \bot$

(SPR). By Definition 4.5 and **SUC** *so* **SPR** *is fulfilled.*

(VAC). As $\mathcal{K} \not\models \bot$, *by Theorem 4.6 there exists an optimal solution* $(\vec{x}^{p*}, \vec{y}^{p*})$, *such that* $\sum_{i=1}^{\bar{b}_\mathcal{K}} y^{p*}_i = 0$. *Since* $\vec{x}^{p*} \in C_r$, $\sum_{j=1}^{\hbar_E} x^{p*}_j = 1$. *By Theorem 4.7 and Definition 4.1,* $\eta^p(\mathcal{K}) = \varphi^E_\mathcal{K}(\vec{x}^{p*}) = \mathcal{K}$.

(IRS). As $\mathcal{K}_1 \stackrel{\wedge}{=} \mathcal{K}_2$, *there exists a bijection* $\alpha : \mathcal{K}_1 \to \mathcal{K}_2$ *such that* $\kappa \equiv \alpha(\kappa)$ *for each* $\kappa \in \mathcal{K}_1$, *that is,* $\mathfrak{V}(\{\kappa\}) = \mathfrak{V}(\{\alpha(\kappa)\})$ *and thus there exists a bijection* $\beta : \eta^p(\mathcal{K}_1) \to \eta^p(\mathcal{K}_2)$ *such that* $\kappa \equiv \beta(\kappa)$ *for each* $\kappa \in \eta^p(\mathcal{K}_1)$. *Therefore,* $\eta^p(\mathcal{K}_1) \stackrel{\wedge}{=} \eta^p(\mathcal{K}_2)$.

(IA). Let $\mathcal{P}_1, \mathcal{P}_2, \mathcal{P}$ *be probability functions such that* $\mathcal{P}_1 \models \mathcal{K}_1$, $\mathcal{P}_2 \models \mathcal{K}_2$, $\mathcal{P} \models \eta^p(\mathcal{K}_1) \cup \eta^p(\mathcal{K}_2)$. *In order to prove* $\eta^p(\mathcal{K}_1) \cup \eta^p(\mathcal{K}_2) \stackrel{\Delta}{=} \eta^p(\mathcal{K}_1 \cup \mathcal{K}_2)$, *it is necessary to show that* $\mathfrak{V}(\eta^p(\mathcal{K}_1) \cup \eta^p(\mathcal{K}_2)) = \mathfrak{V}(\eta^p(\mathcal{K}_1 \cup \mathcal{K}_2))$, *that is,* $\mathcal{P} \in \mathfrak{V}(\eta^p(\mathcal{K}_1) \cup \eta^p(\mathcal{K}_2))$ *iff* $\mathcal{P} \in \mathfrak{V}(\eta^p(\mathcal{K}_1 \cup \mathcal{K}_2))$.

(1) Prove: If $\mathcal{P} \in \mho\left(\eta^p(\mathcal{K}_1) \cup \eta^p(\mathcal{K}_2)\right)$ then $\mathcal{P} \in \mho\left(\eta^p(\mathcal{K}_1 \cup \mathcal{K}_2)\right)$.

As $\eta^p(\mathcal{K}_1) \cup \eta^p(\mathcal{K}_2) \not\models \bot$ so by Definition 3.7 and Definition $\exists \mathcal{P} \in \mho\left(\eta^p(\mathcal{K}_1) \cup \eta^p(\mathcal{K}_2)\right)$. Assume that $\mathcal{P} \notin \mho\left(\eta^p(\mathcal{K}_1 \cup \mathcal{K}_2)\right)$, we have $\mathcal{P} \not\models \eta^p(\mathcal{K}_1 \cup \mathcal{K}_2)$. Then, by Definition 3.5, $\exists \kappa \in \eta^p(\mathcal{K}_1 \cup \mathcal{K}_2) : \mathcal{P} \not\models \kappa$. As $\mathcal{P}_1, \mathcal{P}_2, \mathcal{P}$ lead to the same PKBs in $\mathcal{K}_1 \cup \mathcal{K}_2$ so $\mathcal{P}_1 = \mathcal{P}_2 = \mathcal{P}$. Hence, $\exists \kappa \in \eta^p(\mathcal{K}_1 \cup \mathcal{K}_2) : \mathcal{P} \models \kappa$. This contradicts the hypothesis. Therefore, $\mathcal{P} \in \mho\left(\eta^p(\mathcal{K}_1 \cup \mathcal{K}_2)\right)$.

(2) Prove: If $\mathcal{P} \in \mho\left(\eta^p(\mathcal{K}_1 \cup \mathcal{K}_2)\right)$ then $\mathcal{P} \in \mho\left(\eta^p(\mathcal{K}_1) \cup \eta^p(\mathcal{K}_2)\right)$.

If $\mathcal{P} \in \mho\left(\eta^p(\mathcal{K}_1 \cup \mathcal{K}_2)\right)$ then $\exists \kappa \in \eta^p(\mathcal{K}_1 \cup \mathcal{K}_2) : \mathcal{P} \models \kappa$. As $\eta^p(\mathcal{K}_1) \not\models \bot$ so $\exists \mathcal{P} \models \eta^p(\mathcal{K}_1)$. As $\eta^p(\mathcal{K}_2) \not\models \bot$ so $\exists \mathcal{P} \models \eta^p(\mathcal{K}_2)$. Hence, $\exists \mathcal{P} \models \eta^p(\mathcal{K}_1) \cup \eta^p(\mathcal{K}_2)$. Therefore, $\mathcal{P} \in \mho\left(\eta^p(\mathcal{K}_1) \cup \eta^p(\mathcal{K}_2)\right)$.

*(**WIA**). This property is inferred from **SUC**, **IA**, and Theorem 4.1.*

*(**NOD**). Let $(F \mid G)[\rho_1]$ be any non-tautological probabilistic constraint and $\rho_1 \neq \rho_2$. Consider a PKB $\mathcal{K} = \{(F \mid G)[\rho_1], (F \mid G)[\rho_2], (G)[1]\}$. Firstly, it is easy to see that $\mathcal{K} \models \bot$ and $\mathcal{K} \neq \eta^p(\mathcal{K})$.*

Assume that $(F \mid G)[\rho_1] \in \eta^p(\mathcal{K})$. As the PKB is symmetric in $(F \mid G)[\rho_1]$ and $(F \mid G)[\rho_2]$, we must also have $(F \mid G)[\rho_2] \in \eta^p(\mathcal{K})$.

It follows that $\eta^p(\mathcal{K}) \not\models \bot$ iif $(G)[0] \in \eta^p(\mathcal{K})$. Hence, $\eta^p(\mathcal{K}) = \{(F \mid G)[\rho_1], (F \mid G)[\rho_2], (G)[0]\}$. By Definition 4.5, we have $\vec{\vartheta} = (\rho_1, \rho_2, 0)$. However, by Theorem 4.7, $\eta^p(\mathcal{K}) = \varphi_{\mathcal{K}^*}^{\mathsf{E}}(\sigma_{\mathcal{K}}^{\mathsf{E},p}) = \mathcal{K}^*$ and $\mathcal{K}^* \not\models \bot$. It follows that $\exists \vec{\vartheta}^*$ such that $\eta^p(\mathcal{K}) = \partial_{\mathcal{K}}(\vec{\vartheta}^*)$, where $\forall \vartheta_i \in \vec{\vartheta}^* : \vartheta_i$ could be computed by rules in Theorem 2.1. Hence, $\vec{\vartheta} = \vec{\vartheta}^*$, that is a contradiction. Therefore, $(F \mid G)[\rho_1] \notin \eta^p(\mathcal{K})$. □

4.3.2 The Unnormalized Consistency-Restoring Problem

A model for restoring the unnormalized consistency of a PKB is presented in Figure 4.3.

It consists of the following three steps:

Step 1: Computing the unnormalized inconsistency measure by using Definition 3.31 and Theorem 3.14. If it equals zero then stop. Otherwise, go to step 2.

Step 2: Finding the unnormalized probability vector by using Definition 4.9

Step 3: Calculating new probability for each probabilistic constraint using Definition 4.9.

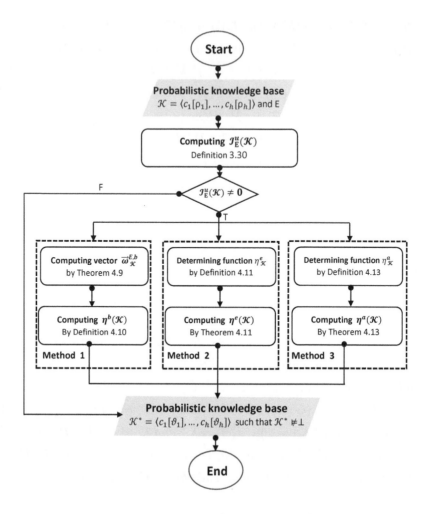

Figure 4.3 A model for restoring the unnormalized consistency.

Below we will present several theorems as the basis for building the consistency-restoring model in Figure 4.3. Theorem 4.9 is employed to find an unnormalized probability vector of a PKB.

Theorem 4.9 ([54]) *Let $\mathcal{R} = \langle \mathcal{B}, \mathbf{E} \rangle$ be a PKB profile. Let $f : \mathbb{R}^{h_E + 2\bar{b}_{\mathcal{K}}} \to \mathbb{R}^*$ such that $f(\vec{\omega}, \vec{\Delta}) = \sum\limits_{i=1}^{\bar{b}_{\mathcal{K}}} (\ell_i + \zeta_i)$. For the following optimization problem:*

$$\arg \min_{(\vec{\omega}, \vec{\Delta}) \in \mathbb{R}^{h_E + 2\bar{b}_{\mathcal{K}}}} \sum_{i=1}^{\bar{b}_{\mathcal{K}}} (\ell_i + \zeta_i) \tag{4.24}$$

subject to $(\vec{\omega}, \vec{\Delta}) \in C_u$

There always exists an optimal solution $\vec{\omega}^*$, simultaneously $\vec{\omega}^*$ is a probability vector of $\mathcal{K} \in \mathcal{B}$ over E.

Proof 4.9 By Theorem 3.14, $\left(\vec{\omega}^*, \vec{\Delta}^*\right)$ is feasible and attains the optimal value f^* so $\vec{\omega}^*$ is a part of the solution of Problem (4.24). Then, we have $\vec{\omega}^* = (\mathcal{P}(\Theta_1), \ldots, \mathcal{P}(\Theta_{\hbar_E}))$. Therefore, by Definition 4.2, $\vec{\omega}$ is a probability vector of $\mathcal{K} \in \mathcal{B}$ over E. □

We call the solution of Problem (4.24) *an unnormalized probability vector* of $\mathcal{K}$ over E, denoted by $\vec{\omega}_{\mathcal{K}}^{\mathsf{E},b}$.

Definition 4.9 is employed to determine a balanced consistency-restoring operator of a PKB.

Definition 4.9 ([51]) Let $\mathcal{R} = \langle \mathcal{B}, \mathsf{E} \rangle$ be a PKB profile A balanced consistency-restoring operator $\eta^b : \mathbb{K} \to \mathbb{K}$ of $\mathcal{K} \in \mathcal{B}$ is defined as follows:

$$\eta^b(\mathcal{K}) = \partial_{\mathcal{K}}(\vec{\vartheta}) \qquad (4.25)$$

where $\vec{\vartheta} = (\vartheta_1, \ldots, \vartheta_{\bar{b}_{\mathcal{K}}})$ with $\vartheta_i = |\rho_i + \ell_i^* - \zeta_i^*|$ such that $\ell_i^*, \zeta_i^* \in \vec{\Delta}^*$.

Theorem 4.10 presents the relationship between the balanced consistency-restoring operator and desirable properties.

Theorem 4.10 ([54]) The balanced consistency-restoring operator η^b satisfies *SPR, SUC, VAC, WIA, IA, IRS,* and *NOD.*

Proof 4.10 *(SUC).* By Definition 4.5, $\eta^b(\mathcal{K}) = \partial_{\mathcal{K}}(\vartheta_1, \ldots, \vartheta_{\bar{b}_{\mathcal{K}}}) = \langle (F_1 | G_1)[\vartheta_1], \ldots, (F_{\bar{b}_{\mathcal{K}}} | G_{\bar{b}_{\mathcal{K}}})[\vartheta_{\bar{b}_{\mathcal{K}}}] \rangle$. As $\vartheta_i = |\rho_i + l_i^* - \zeta_i^*|$ and where l_i^*, ζ_i^* is an optimal solution (4.24) with $i = \overline{1, \bar{b}_{\mathcal{K}}}$. It is easy to see that $\mathcal{I}_{\mathcal{K}}^u = 0$ so $\eta(\mathcal{K}) \not\models \perp$.

(SPR). By Definition 4.9 and *SUC* so *SPR* is fulfilled.

(VAC). As $\mathcal{K} \not\models \perp$ and by Definition 3.31, $\vec{\Delta}^* = 0$, that is, $l_i^* = 0, \zeta_i^* = 0$, $\forall i = \overline{1, \bar{b}_{\mathcal{K}}}$. Therefore, $\eta^b(\mathcal{K}) = \partial_{\mathcal{K}}\left(|\rho_1 + l_1^* - \zeta_1^*|, \ldots, |\rho_{\bar{b}_{\mathcal{K}}} + l_{\bar{b}_{\mathcal{K}}}^* - \zeta_{\bar{b}_{\mathcal{K}}}^*|\right) = \partial_{\mathcal{K}}\left(\rho_1, \ldots, \rho_{\bar{b}_{\mathcal{K}}}\right) = \mathcal{K}$.

(NOD). Let $(F | G)[\rho_1]$ be any non-tautological probabilistic constraint and $\rho_1 \neq \rho_2$. Consider a PKB $\mathcal{K} = \{(F | G)[\rho_1], (F | G)[\rho_2], (G)[1]\}$. Firstly, it is easy to see that $\mathcal{K} \models \perp$ and $\mathcal{K} \neq \eta^b(\mathcal{K})$. Assume that $(F | G)[\rho_1] \in \eta^b(\mathcal{K})$. As the PKB is symmetric in $(F | G)[\rho_1]$ and $(F | G)[\rho_2]$, we must also have $(F | G)[\rho_2] \in \eta^b(\mathcal{K})$. In order to make $\eta^b(\mathcal{K})$ consistent, it follows that $(G)[0] \in \eta^b(\mathcal{K})$. We have

$\eta^b(\mathcal{K}) = \{(F\,|G)\,[\rho_1]\,,(F\,|G)\,[\rho_2]\,,(G)\,[0]\}$ so $\vec{\vartheta} = (\rho_1,\rho_2,0)$. However, by Definition 4.9, $\eta^b(\mathcal{K}) = \{(F\,|G)\,[|\rho_1 + l_1^* - \zeta_1^*|]\,,(F\,|G)\,[|\rho_2 + l_2^* - \zeta_2^*|]\,,(G)\,[|l_3^* - \zeta_3^*|]\}$ and thus $\vec{\vartheta}^* = (|\rho_1 + l_1^* - \zeta_1^*|,|\rho_2 + l_2^* - \zeta_2^*|,|l_3^* - \zeta_3^*|)$. However, as $\mathcal{K} \models \perp$ and by Theorem 4.10, $\exists l_i^* \neq 0, \zeta_i^* \neq 0$. Hence, $\vec{\vartheta} \neq \vec{\vartheta}^*$, that is a contradiction. Therefore, $(F\,|G)\,[\rho_1] \notin \eta^b(\mathcal{K})$.

These properties **IA**, **WIA**, and **IRS** could be proved in a similar way as Theorem 4.8. $\qquad\square$

Example 4.5 *Let's continue example 3.13. By Theorem 4.9, $\vec{\Delta}$ is computed by solving the optimization problem:*

$$\arg\min_{(\vec{\omega},\vec{\Delta})\in\mathbb{R}^{8+10}} (\ell_1 + \ell_2 + \ell_3 + \ell_4 + \ell_5 + \zeta_1 + \zeta_2 + \zeta_3 + \zeta_4 + \zeta_5) \qquad (4.26)$$

subject to (3.61)-(3.69).
    Thus, $\vec{\omega}_{\mathcal{K}}^{E,b^*} = (0,0.35,0.29,0.06,0,0,0.16,0.14))$, $\vec{\Delta}^* = (\vec{\ell}^*,\vec{\zeta}^*)$
    where
    $\vec{\ell}^* = (0,0.05,0,0,0)$,
    $\vec{\zeta}^* = (0,0,0,0,0)$.
    By Definition 4.9, $\vec{\vartheta} = (0.7,0.35,0,45,0.5,0.64)$.
    Then, $\eta^b(\mathcal{K}_1) = \{(H)[0.7],(T)[0.35],(D)[0.45],(T\,|H)[0.5],(H\,|D)[0.64]\}$.
    New probability values of probabilistic constraints in $\mathcal{K}_1$ and $\mathcal{K}_5$ are shown in Table 4.7 after the operator η^b is employed.

Definition 4.10 ([77]) *Let $\mathcal{K} = \{c_1[\rho_1],\dots,c_{\bar{b}_\mathcal{K}}[\rho_{\bar{b}_\mathcal{K}}]\}$ be a PKB. An equitable deformation function of $\mathcal{K}$ is function $\Gamma_{\mathcal{K}}^e : \mathbb{R}_{[0,1]} \to \mathbb{K}$ defined as follows:*

$$\Gamma_{\mathcal{K}}^e(\Psi) = \partial_{\mathcal{K}}(\vec{\vartheta}^e) \qquad (4.27)$$

where $\vec{\vartheta}^e = (\vartheta_1^e,\dots,\vartheta_{\bar{b}_\mathcal{K}}^e)$ with $\vartheta_i^e = \rho_i + (0.5 - \rho_i)\Psi$ and $\Psi \in \mathbb{R}_{[0,1]}$.

Theorem 4.11 is employed to determine an equitable consistency-restoring operator of a PKB.

Theorem 4.11 ([77]) *Let $\mathcal{D} = \langle \mathcal{B}, \mathsf{E} \rangle$ be a PKB profile and $\mathcal{K} \in \mathcal{B}$. Let $\eta^e : \mathbb{K} \to \mathbb{K}$ be an equitable consistency-restoring operator of $\mathcal{K}$. There exists $\Psi^* = \min\{\Psi \in \mathbb{R}_{[0,1]} : \mathcal{I}(\Gamma_{\mathcal{K}}^e(\Psi)) = 0\}$ such that*

$$\eta^e(\mathcal{K}) = \Gamma_{\mathcal{K}}^e(\Psi^*) \qquad (4.28)$$

Proof 4.11 *If there exists $\Psi^* = \min\{\Psi \in \mathbb{R}_{[0,1]} : \mathcal{I}(\Gamma^e_{\mathcal{K}}(\Psi)) = 0\}$ then $\Gamma^e_{\mathcal{K}}(\Psi) \not\models \bot$.*

By Definition 4.10, there exists $\vec{\vartheta}^{e*} = (\vartheta^{e*}_1, \ldots, \vartheta^{e*}_{\bar{b}_{\mathcal{K}}})$ with $\vartheta^{e*}_i = \rho_i + (0.5 - \rho_i)\Psi^*$ such that $\eta^e(\mathcal{K}) = \partial_{\mathcal{K}}(\vec{\vartheta}^{e*})$. Since $\Gamma^e_{\mathcal{K}}(\Psi) \not\models \bot$, it follows that $\partial_{\mathcal{K}}(\vec{\vartheta}^{e*}) \not\models \bot$. Therefore, $\eta^e(\mathcal{K}) \not\models \bot$. □

Theorem 4.12 presents the relationship between the equitable consistency-restoring operator and desirable properties.

Theorem 4.12 ([77]) *The equitable consistency-restoring operator η^e satisfies* SUC, SPR, VAC, IRS, WIA, IA.

Proof 4.12 *The proof is similar to that of Theorem 4.10*

Example 4.6 *Consider $\mathcal{K}_1$ in example 2.3.*

The smallest of Ψ such that $\mathcal{I}(\eta^e(\mathcal{K}_1)) = 0$ is $\Psi^* = 0.167$.

Then, $\eta^e(\mathcal{K}_1) = \{(H)[0.67], (T)[0.33], (D)[0.46], (T|H)[0.5], (H|D)[0.62]\}$.

New probability values of probabilistic constraints in $\mathcal{K}_1$ and $\mathcal{K}_5$ are shown in Table 4.7 after the operator η^e is employed.

Definition 4.11 ([77]) *Let $\mathcal{K} = \{c_1[\rho_1], \ldots, (c_{\bar{b}_{\mathcal{K}}})[\rho_{\bar{b}_{\mathcal{K}}}]\}$. For each $\kappa_i \in \mathcal{K}\ \forall i = \overline{1, \bar{b}_{\mathcal{K}}}$,*

- *A mean amerced distance measure of κ_i is defined by $d(\kappa_i) = |\ell_i - \zeta_i|$*
- *The sign of κ_i is defined as follows:*

$$s(\kappa_i) = \begin{cases} -1 & if\ \ell_i - \zeta_i < 0 \\ 0 & if\ \ell_i - \zeta_i = 0 \\ 1 & otherwise \end{cases}$$

where $\ell_i \in \vec{\ell}^*$ and $\zeta_i \in \vec{\zeta}^*$ with $\vec{\Delta}^* = (\vec{\ell}^*, \vec{\zeta}^*)$ is a part of the optimal solution of problem (4.24).

- *The normalized amerced vector of $\mathcal{K}$ is defined by $\vec{\delta} = (\delta_1, \ldots, \delta_{\bar{b}_{\mathcal{K}}})$,*

where

$\delta_i = \frac{s(\kappa_i)d(\kappa_i)}{\widehat{\delta}}$ *and*

$\widehat{\delta} = \min\{|s(\kappa_i)d(\kappa_i)| : s(\kappa_i)d(\kappa_i) \neq 0\ \forall i = \overline{1, \bar{b}_{\mathcal{K}}}\}$.

If $s(\kappa_i)d(\kappa_i) = 0\ \forall i = \overline{1, \bar{b}_{\mathcal{K}}}$, $\widehat{\delta} = 0$.

In order to ensure that the values of probabilistic constraints always satisfy the basic principle of probability $0 \leq \mathcal{P}(A) \leq 1$ for all $A \in \mathsf{E}$, the probabilistic principle fulfilled function is used. It is function $p : \mathbb{R} \to \mathbb{R}_{[0,1]}$ defined as follows:

$$p(\rho) = \begin{cases} \rho & if \ \rho \in [0,1] \\ 1 & \rho > 1 \\ 0 & otherwise \end{cases}$$

Definition 4.12 ([77]) *Let* $\mathcal{K} = \{c_1[\rho_1], \ldots, \left(c_{\bar{b}_\mathcal{K}}\right)[\rho_{\bar{b}_\mathcal{K}}]\}$. *An amerced deformation function of* $\mathcal{K}$ *is function* $\Gamma^a_\mathcal{K} : \mathbb{R}_{[0,1]} \to \mathbb{K}$ *defined as follows:*

$$\Gamma^a_\mathcal{K}(\Psi) = \partial_\mathcal{K}(\vec{\vartheta}^a) \tag{4.29}$$

where
$$\vec{\vartheta}^a = (\vartheta^a_1, \ldots, \vartheta^a_{\bar{b}_\mathcal{K}})$$
with $\vartheta^a_i = p(\rho_i + \Psi\delta_i)$ *and* $\Psi \in \mathbb{R}_{[0,1]}$.

Theorem 4.13 is employed to determine an amerced consistency-restoring operator of a PKB.

Theorem 4.13 ([77]) *Let* $\mathcal{D} = \langle \mathcal{B}, \mathsf{E} \rangle$ *be a PKB profile and* $\mathcal{K} \in \mathcal{B}$ *be a PKB. Let* $\eta^a : \mathbb{K} \to \mathbb{K}$ *be an amerced consistency-restoring operator of* $\mathcal{K}$. *There exists* $\Psi^* = \min\{\Psi \in \mathbb{R}_{[0,1]} : \mathcal{I}(\Gamma^a_\mathcal{K}(\Psi)) = 0\}$ *such that*

$$\eta^a(\mathcal{K}) = \Gamma^a_\mathcal{K}(\Psi^*) \tag{4.30}$$

Proof 4.13 *The proof is a modification of the proof of Theorem 4.11 for a amerced deformation function of* $\mathcal{K}$.

Theorem 4.14 presents the relationship between the amerced consistency-restoring operator and desirable properties.

Theorem 4.14 ([77]) *The amerced consistency-restoring operator* $\eta^a(\mathcal{K})$ *satisfies* SUC, SPR, NOD, WIA *and* IA.

Proof 4.14 *The proof is similar to that of Theorem 4.10*

Example 4.7 *Let's continue example 4.5. We have* $\vec{\Delta}^* = (\vec{\ell}^*, \vec{\zeta}^*)$ *where* $\vec{\ell}^* = (0, 0.05, 0, 0, 0), \vec{\zeta}^* = (0, 0, 0, 0, 0)$. *Table 4.6 presents the normalized amerced vector of* $\mathcal{K}_1$.

Table 4.6 The normalized amerced vector of $\mathcal{K}_1$

κ_i	$d(\kappa_i)$	$s(\kappa_i)$	$s(\kappa_i)d(\kappa_i)$	$\vec{\delta}$
$(H)[0.7]$	0	0	0	0
$(T)[0.3]$	0.05	1	0.05	1
$(D)[0.45]$	0	0	0	0
$(T\,|\,H\,)\,[0.5]$	0	0	0	0
$(H\,|\,D\,)\,[0.64]$	0	0	0	0

We have $\widehat{\delta} = 0.05$. Therefore, $\vec{\delta} = (0,1,0,0,0)$

Then, $\eta^a(\mathcal{K}_1) = \{(H)[0.7], (T)[0.35], (D)[0.45], (T\,|\,H\,)[0.5], (H\,|\,D)[0.64]\}$.

New probability values of probabilistic constraints in $\mathcal{K}_1$ and $\mathcal{K}_5$ are shown in Table 4.7 after the operator η^a is employed.

Table 4.7 New probability values of probabilistic constraints in $\mathcal{K}_1$ and $\mathcal{K}_5$ with respect to the unnormalized consistency-restoring operators

$\kappa_i \in \mathcal{K}_1$	$\eta^b(\mathcal{K}_1)$	$\eta^e(\mathcal{K}_1)$	$\eta^a(\mathcal{K}_1)$	$\kappa_i \in \mathcal{K}_5$	$\eta^b(\mathcal{K}_5)$	$\eta^e(\mathcal{K}_5)$	$\eta^a(\mathcal{K}_5)$
$(H)[0.7]$	0.7	0.67	0.7	$(H)[0.8]$	0.8	0.77	0.8
$(T)[0.3]$	0.35	0.33	0.35	$(T)[0.31]$	0.37	0.33	0.336
$(D)[0.45]$	0.45	0.46	0.45	$(D)[0.5]$	0.5	0.51	0.5
$(T\,|\,H\,)\,[0.5]$	0.5	0.5	0.5	$(T\,|\,H\,)\,[0.42]$	0.42	0.43	0.42
$(H\,|\,D\,)\,[0.64]$	0.64	0.62	0.64	$(H\,|\,D\,)\,[0.6]$	0.6	0.59	0.6

The relationship between consistency-restoring operators and desirable properties is summarized in Table 4.8. It shows that operators η^p, η^∞, and η^b seem to be attractive when compared to other operators hitherto considered because they satisfy all properties. Operators η^1, η^e also satisfy all properties except the property NOD. The operator η^a has its weakness in terms of the properties of Irrelevance of Syntax and Vacuity

Table 4.8 The relationship between consistency-restoring operators and desirable properties

Operators	SUC	SPR	VAC	IRS	NOD	WIA	IA	Theorem
η^1	√	√	√	√	-	√	√	4.8
η^p	√	√	√	√	√	√	√	4.8
η^∞	√	√	√	√	√	√	√	4.8
η^b	√	√	√	√	√	√	√	4.10
η^e	√	√	√	√	-	√	√	4.12
η^a	√	√	-	-	√	√	√	4.14

4.4 ALGORITHMS FOR RESTORING CONSISTENCY

The following algorithm is employed to find the new probability value of probabilistic constraints (**FPVPC**) after the integration process and the consistency-restoring process.

Algorithm FPVPC:

Input: A satisfying restored probability vector or a satisfying probability vector over a set of events.

Output: A consistent PKB.

Idea: :For each probabilistic constraint, the new probability value is computed by using probability rules.

Algorithm 4 Computing the new probability value of probabilistic constraints (FPVPC) [51]

Input : $\langle \vec{\omega}_{\mathcal{K}}^{\mathsf{E}}, \mathsf{E} \rangle$
Output: $\mathcal{K}$ with $\mathcal{K} \not\models \bot$

1 Function FPVPC($\vec{\omega}_{\mathcal{K}}^{\mathsf{E}}, \mathbf{E}$)
  begin
2 | $Fcc = \text{FindingSCC}(\mathsf{E})$;
3 | for $c[\rho] \in \mathcal{K}$ do
4 | | $x \leftarrow 0$;
5 | | for $cc \in Fcc$ do
6 | | | if $App(c[\rho]) \subseteq App(cc)$ then $x \leftarrow x + \text{getValue}(\vec{\omega}_{\mathcal{K}}^{\mathsf{E}}[cc])$;
7 | | end for
8 | | $\rho \leftarrow x$;
9 | end for
10 | for $c[\rho] \in \mathcal{K}$ do
11 | | if $c = (F|G)$ then $\rho \leftarrow \rho/\text{getValue}((G)[\rho])$;
12 | end for
13 | return $\mathcal{K}$;
14 end

Algorithm 4 is built on the basis of Theorem 2.1 and consists of two stages:

(i) Computing the new probability values of probabilistic constraints $(F)[\rho]$ $(F)[\rho]$ (from line 3 to 9)

(ii) Computing the new probability values of probabilistic constraints $(F|G)[\rho]$ (from line 10 to 12).

Theorem 4.15 is employed to evaluate the complexity of Algorithm 4 (**FPVPC**) for calculating inconsistency measures.

Theorem 4.15 ([51]) *Let $\mathcal{R} = \langle \mathcal{B}, \mathsf{E} \rangle$ be PKB profile and, $\mathcal{K} \in \mathcal{B}$. The complexity of Algorithm 4 (FPVPC) is*

$$\mathcal{O}(\bar{b}_{\mathcal{K}} \times 3^{\bar{b}_{\mathcal{K}}})$$

Proof 4.15 *The cost of the stage 1 is $\mathcal{O}\left(\bar{b}_{\mathcal{K}} \times \hbar_{\mathsf{E}}\right)$. The cost of the stage 2 is $\mathcal{O}\left(\bar{b}_{\mathcal{K}}\right)$. Therefore, the complexity of Algorithm 4 (FPVPC) is $\mathcal{O}\left(max\left\{\bar{b}_{\mathcal{K}} \times \hbar_{\mathsf{E}}, \bar{b}_{\mathcal{K}}\right\}\right) = \mathcal{O}\left(\bar{b}_{\mathcal{K}} \times \hbar_{\mathsf{E}}\right) = \mathcal{O}(\bar{b}_{\mathcal{K}} \times 3^{\bar{b}_{\mathcal{K}}})$.* □

Algorithm 5 (**RCK**-*Restoring the Consistency of a PKB*) is employed to solve the norm-based consistency-restoring problem.

Algorithm **RCK**:

Input: An inconsistent PKB, a set of events, and pu-norm to identify the problem type

Output: A consistent PKB.

Idea: Build and solve the optimization problem with the objective function based on the principle of maximum entropy, the constraints based on probability rules and an inconsistency measure.

Algorithm 5 Restoring the consistency of a PKB (RCK) [51]

Input : $\langle \mathcal{K}, \mathsf{E}, pu \rangle$
Output: $\mathcal{K}^*$ with $\mathcal{I}(\mathcal{K}^*) = 0$

1 Function RCK($\mathcal{K}, \mathsf{E}, pu$)
   begin
2 | $\epsilon = \text{CIM}(\mathcal{K}, \mathsf{E}, pu)$;
3 | $pr = \text{OptimizationProblem}(MINIMIZE)$;$Fcc = \text{FindingSCC(E)}$;
4 | $sx.\text{setEmpty}()$; $sy.\text{setEmpty}()$;$g.\text{setEmpty}()$;
5 | **for** $cc \in Fcc$ **do**
6 | | $pr.\text{addCs}(\langle x[cc] \geq 0 \rangle)$; $sx.\text{addCVar}(\langle x[cc] \rangle)$;
7 | | $g.\text{addCVar}(\langle x[cc] \cdot log(x[cc]) \rangle)$;
8 | **end for**
9 | $pr.\text{addCs}(\langle sx = 1 \rangle)$;
10 | **for** $c[\rho] \in \mathcal{K}$ **do** $pr.\text{addCs}(\langle y[c] \geq 0 \rangle)$; $sy.\text{addCVar}(\langle y[c] \rangle)$;
11 | $pr.\text{addCs}(\langle sy = \epsilon \rangle)$;
12 | **for** $c[\rho] \in \mathcal{K}$ **do**
13 | | $lie.\text{setEmpty}()$; $rie.\text{setEmpty}()$;
14 | | **for** $cc \in Fcc$ **do**
15 | | | **if** $App(c[\rho]) \subseteq App(cc)$ **then** $lie.\text{addCVar}(\langle x[cc] \rangle)$;
16 | | **end for**
17 | | $rie.\text{addCVar}(\langle y[c] + \rho \rangle)$; $pr.\text{addCs}(\langle lie = rie \rangle)$;
18 | **end for**
19 | $sol = \text{OpenOpt.solve}(NLP, pr.\text{addOF}(\langle g \rangle))$;
20 | $\mathcal{K}^* = \text{FPVPC}(\text{getOptimalSolution}(sol), \mathsf{E})$;
21 | **return** $\mathcal{K}^*$;
22 **end**

Algorithm 5 is based on Theorem 4.6 and consists of three stages:

(i) Computing the inconsistency measure of $\mathcal{K}$ with respect to a certain p-norm (line 2) by using Algorithm 3.

(ii) Building the optimization problem (from line 3 to 19) to find the satisfying restored probability vector of a PKB.

- From line 4 to 9, building constraints $\sum_{j=1}^{\hbar_E} x_j = 1$ and $\vec{x} \geq \vec{0}$
- Line 7, building the objective function $g(x,y) = \sum_{j=1}^{\hbar_E} x_i \cdot log(x_i)$.
- From line 10 to 11, building constraintsc $\sum_{i=1}^{\bar{b}_\mathcal{K}} y_i = \epsilon^p$ and $\vec{y} \geq 0$.
- From line 12 to 18, building constraints $C_\mathcal{K}^{\mathsf{E},+} \cdot \vec{x} = \vec{y} + z_\mathcal{K}^{\mathsf{E}}$.
- Line 19, solving the optimization problem.

(iii) Computing the new probability value of probabilistic constraints (line 20).

Theorem 4.16 evaluates the complexity of Algorithm 5 (**RCK**) for restoring the consistency of a PKB.

Theorem 4.16 ([51]) *Let $\mathcal{R} = \langle \mathcal{B}, \mathsf{E} \rangle$ be PKB profile and $\mathcal{K} \in \mathcal{B}$. The complexity of Algorithm 5 (**RCK**) is*

$$\mathcal{O}(\bar{b}_\mathcal{K}^2 \times 3^{\bar{b}_\mathcal{K}})$$

Proof 4.16 - *In stage 1, by Theorem 3.16, the cost for computing inconsistency measures:*

- *$\mathcal{O}(\bar{b}_\mathcal{K} \times 3^{\bar{b}_\mathcal{K}})$ with respect to 1-norm and ∞-norm*
- *$\mathcal{O}(\bar{b}_\mathcal{K}^2 \times 3^{\bar{b}_\mathcal{K}})$ with respect to p-norm ($p > 1$ and $p \neq \infty$), and unnormalized form.*
- *In stage 2, by Theorem 4.6, the objective function is non-linear. The satisfying probability vector of a consistent PKB $\mathcal{K}$ is the optimal solution of the problem (4.14) that is solved by interior-point method [56]. By Problem (3.70) with constraints (3.71-3.72) and by Problem (3.25), $f_0(\vec{\omega}) = \sum_{i=1}^{\hbar_E} \omega_i log(\omega_i)$, $f_i(\vec{\omega}) = -\omega_i \forall i = \overline{1, \hbar_E}$, $Q = \underbrace{(1, \ldots, 1)}_{\hbar_E}$. By problem (3.73), $\nabla \Phi(\vec{\omega}) = \sum_{i=1}^{\hbar_E} \frac{1}{\omega_i}$. Hence, $t_{RC}^{(0)}$ is the minimized value $\inf_{\mathcal{G}} \|t \nabla f_0(\vec{\omega}^{(0)}) + \nabla \Phi(\vec{\omega}^{(0)}) + Q^T \mathcal{G}\|_2$. Let N_{RC} be the cost for solving optimization Problem (4.14). By Problem (3.75) and Problem (4.4),*

$N_{RC} = \frac{\log \frac{\bar{b}_\mathcal{K} + \hbar_E}{t_{RC}^{(0)} \varepsilon}}{\log \mu} \left(\frac{(\bar{b}_\mathcal{K} + \hbar_E)(\mu - 1 - \log \mu)}{\gamma} + c \right).$

Let $g(n) = N_{RC}$ with $n = \bar{b}_\mathcal{K}$. Then, $g(n) = \frac{\log(n+2^n) - \log t}{\log(1 + \frac{1}{\sqrt{n+2^n}})} (\frac{1}{\gamma}(n + 2^n)(\frac{1}{\sqrt{n+2^n}} - \log(1 + \frac{1}{\sqrt{n+2^n}})) + c$ where t, γ, c are constants. The fact that $g(n)$ is $\mathcal{O}(n^2 \times 33^n)$. Therefore, the cost for stage 2 is $\mathcal{O}(\bar{b}_\mathcal{K}^2 \times 3^{\bar{b}_\mathcal{K}})$.

- *In stage 3, by Theorem 4.15, the cost is $\mathcal{O}(\bar{b}_\mathcal{K} \times 3^{\bar{b}_\mathcal{K}})$.*

Therefore, the complexity of Algorithm 5 (RCK) is $\mathcal{O}(\bar{b}_{\mathcal{K}}^2 \times 3^{\bar{b}_{\mathcal{K}}})$. □

The following algorithm is employed to restore the consistency of a PKB by employing a balanced consistency-restoring operator (RCPB).

Algorithm RCPB:

Input: An inconsistent PKB, a set of event.

Output: A consistent PKB.

Idea: Finding a balanced consistency-restoring operator.

Algorithm 6 Inconsistency resolution by employing a balanced consistency-restoring operator (RCPB)

Input : $\langle \mathcal{K}, \mathsf{E} \rangle$

Output: $\mathcal{K}^*$ with $\mathcal{I}(\mathcal{K}^*) = 0$

---

1 Function RCPB($\mathcal{K}, \mathsf{E}$)

   begin

2     $\vec{\Delta}^* = (\vec{\ell}^*, \vec{\zeta}^*) = \texttt{getOptimalSolution}(\texttt{UBOP}(\mathcal{K}, \mathsf{E}))$;

3   **for** $c_i[\rho_i] \in \mathcal{K}$ *and* $\ell_i^* \in \vec{\ell}^*$ *and* $\zeta_i^* \in \vec{\zeta}^*$ **do** $\vartheta_i^b = |\rho_i + \ell^*{}_i - \zeta^*{}_i|$;

4     $\eta^b(\mathcal{K}) = \partial_{\mathcal{K}}(\vartheta_1^b, \dots, \vartheta_{b_{\mathcal{K}}}^b)$;

5     $\mathcal{K}^* = \eta^b(\mathcal{K})$;

6   **return** $\mathcal{K}^*$;

7 **end**

Algorithm 6 is based on Definition 4.9. It consists of two stages:

(i) Solving the optimization problem (line 2) by employing Algorithm 2 (UNOP).

(ii) From line 3 to 5 finding a consistent PKB.

Theorem 4.17 evaluates the complexity of Algorithm 6 (RCPB) for restoring the consistency of a PKB.

Theorem 4.17 ([51]) *Let $\mathcal{R} = \langle \mathcal{B}, \mathsf{E} \rangle$ be PKB profile and $\mathcal{K} \in \mathcal{B}$. The complexity of Algorithm 6 (RCPB) is*

$$\mathcal{O}(\bar{b}_{\mathcal{K}}^2 \times 3^{\bar{b}_{\mathcal{K}}})$$

Proof 4.17 *By Theorem 3.16, the cost of stage 1 is $\mathcal{O}(\bar{b}_{\mathcal{K}}^2 \times 3^{\bar{b}_{\mathcal{K}}})$. In the second stage, the cost for finding $\eta^b(\mathcal{K})$ is $\mathcal{O}(\bar{b}_{\mathcal{K}})$. Therefore, the complexity of Algorithm 6 (RCPB) is $\mathcal{O}(\bar{b}_{\mathcal{K}}^2 \times 3^{\bar{b}_{\mathcal{K}}})$.* □

The following algorithm is employed to restore the consistency of a PKB by employing the equitable consistency-restoring operator (RCPE).

Algorithm **RCPE**:

Input: An inconsistent PKB, a set of events.

Output: A consistent PKB.

Idea: Finding an equitable consistency-restoring operator.

Algorithm 7 Inconsistency resolution by employing the equitable consistency-restoring operator (RCPE)

Input : $\langle \mathcal{K}, \mathsf{E} \rangle$
Output: $\mathcal{K}^*$ with $\mathcal{I}(\mathcal{K}^*) = 0$

1 Function RCPE($\mathcal{K}, \mathsf{E}$)
  begin
2     $\epsilon = 10^{-3}$; $\Psi = 0$;
3   **while** $\Psi < 1$ *and* $\mathcal{I}(\eta^e(\mathcal{K})) \neq 0$ **do**
4       $\Psi = \Psi + \epsilon$;
5   **for** $c_i[\rho_i] \in \mathcal{K}$ **do** $\vartheta_i^e = \rho_i + (0.5 - \rho_i)\Psi$;
6       $\eta^e(\mathcal{K}) = \partial_{\mathcal{K}}(\vartheta_1^e, \dots, \vartheta_{\bar{b}_{\mathcal{K}}}^e)$;
7   **end while**
8     $\mathcal{K}^* = \eta^e(\mathcal{K})$;
9   **return** $\mathcal{K}^*$;
10 **end**

Algorithm 7 is based on Definition 4.10 and Theorem 4.11. It consists of two stages:

(i) Computing the inconsistency measure of new PKB (line 3) by employing Algorithm 3. (**CIM**)

(ii) Finding an equitable deformation function $\eta^e(\mathcal{K}) = \Gamma_{\mathcal{K}}^e = \partial_{\mathcal{K}}(\vartheta_1^e, \dots, \vartheta_n^e)$ of $\mathcal{K}$ (from line 3 to 7)

The following theorem evaluates the complexity of Algorithm 7 (**RCPE**) for restoring the consistency of a PKB.

Theorem 4.18 ([77]) *Let* $\mathcal{R} = \langle \mathcal{B}, \mathsf{E} \rangle$ *be the PKB profile and* $\mathcal{K} \in \mathcal{B}$. *The computational complexity of the algorithm* RCPE *is* $\mathcal{O}\left(\bar{b}_{\mathcal{K}}^3 \times 3^{\bar{b}_{\mathcal{K}}}\right)$.

Proof 4.18 *By Theorem 3.16, the cost of stage 1 is* $\mathcal{O}(\bar{b}_{\mathcal{K}}^2 \times 3^{\bar{b}_{\mathcal{K}}})$. *In the second stage, the cost for finding an equitable deformation function* $\Gamma_{\mathcal{K}}^e$ *is* $\mathcal{O}(\bar{b}_{\mathcal{K}})$. *Therefore, the cost of the algorithm* RCPE *is* $\mathcal{O}\left(\bar{b}_{\mathcal{K}}^3 \times 3^{\bar{b}_{\mathcal{K}}}\right)$. $\qquad\square$

The following algorithm is employed to restore the consistency of a PKB by employing an amerced consistency-restoring operator (**RCPA**).

Algorithm **RCPA**:

Input: An inconsistent PKB, a set of events.

Output: A consistent PKB.

Idea: Finding an amerced consistency-restoring operator.

Algorithm 8 Inconsistency resolution by employing the amerced consistency-restoring operator (RCPA)

Input : $\langle \mathcal{K}, \mathsf{E} \rangle$
Output: $\mathcal{K}^*$ with $\mathcal{I}(\mathcal{K}^*) = 0$

1 Function RCPA($\mathcal{K}, \mathsf{E}$)
  begin
2 | $\vec{\Delta}^* = (\vec{\ell}^*, \vec{\zeta}^*) = \texttt{getOptimalSolution}(\texttt{UBOP}(\mathcal{K}, \mathsf{E}))$;
3 | for $c_i[\rho_i] \in \mathcal{K}$ and $\ell_i^* \in \vec{\ell}^*$ and $\zeta_i^* \in \vec{\zeta}^*$ do
4 | | $d(c_i[\rho_i]) = \ell_i - \zeta_i$;
5 | | if $d(c_i[\rho_i]) < 0$ then $s(c_i[\rho_i]) = -1$;
6 | | else if $d(c_i[\rho_i]) = 0$ then $s(c_i[\rho_i]) = 0$;
7 | | else $s(c_i[\rho_i]) = 1$;
8 | | $\delta_i = s(c_i[\rho_i]) * d(c_i[\rho_i])$;
9 | end for
10 | if $\forall c_i[\rho_i] \in \mathcal{K} : \delta_i = 0$ then $\widehat{\delta} = 0$;
11 | else
12 | | for $c_i[\rho_i] \in \mathcal{K}$ do
13 | | | if $\delta_i \neq 0$ and $\widehat{\delta} > \delta_i$ then $\widehat{\delta} = \delta_i$;
14 | | end for
15 | end if
16 | for $c_i[\rho_i] \in \mathcal{K}$ do $\delta_i = \frac{\delta_i}{\widehat{\delta}}$
17 | $\epsilon = 10^{-3}$; $\Delta = 0$;
18 | while $\Delta < 1$ and $\mathcal{I}(\eta^a(\mathcal{K})) \neq 0$ do
19 | | $\Delta = \Delta + \epsilon$;
20 | | for $c_i[\rho_i] \in \mathcal{K}$ do
21 | | | $p_i = \rho_i + \Delta \delta_i$;
22 | | | if $p_i < 0$ then $\vartheta_i^a = 0$;
23 | | | else if $p_i > 1$ then $\vartheta_i^a = 1$;
24 | | | else $\vartheta_i^a = p_i$;
25 | | end for
26 | | $\eta^a(\mathcal{K}) = \partial_{\mathcal{K}}(\vartheta_1^a, \ldots, \vartheta_{b_{\mathcal{K}}}^a)$;
27 | end while
28 | $\mathcal{K}^* = \eta^a(\mathcal{K})$;
29 | return $\mathcal{K}^*$;
30 end

Algorithm 8 is based on Definition 4.12 and Theorem 4.13. It consists of three stages:

(i) Solving the optimization problem (line 2) by employing Algorithm 2 (**UNOP**).

(ii) Computing the inconsistency measure of the new PKB (line 18) by employing Algorithm 3 (**CIM**).

(iii) Finding an amerced deformation function $\eta^a(\mathcal{K}) = \Gamma_{\mathcal{K}}^a = \partial_{\mathcal{K}}(\vartheta_1^a, \ldots, \vartheta_n^a)$ of $\mathcal{K}$ (from line 18 to 27).

The following theorem evaluates the complexity of Algorithm 8 (**alg:RCPA**) for restoring the consistency of a PKB.

Theorem 4.19 ([**77**]) *Let* $\mathcal{R} = \langle \mathcal{B}, \mathsf{E} \rangle$ *be PKB profile and* $\mathcal{K} \in \mathcal{B}$. *The computational complexity of Algorithm* RCPA *is* $\mathcal{O}\left(\bar{b}_{\mathcal{K}}^3 \times 3^{\bar{b}_{\mathcal{K}}}\right)$.

Proof 4.19 *The cost for finding the mean amerced vector and the cost of the second stage is similar to the cost of the first stage in Theorem 4.18. Therefore, the cost of* RCPA *is* $\mathcal{O}\left(\bar{b}_{\mathcal{K}}^3 \times 3^{\bar{b}_{\mathcal{K}}}\right)$. □

4.5 CONCLUDING REMARKS

In this chapter, we have synthesized the methods for solving inconsistencies of knowledge bases. We have focused on the method of changing the probability values of knowledge bases on building a general model for restoring consistency. Our model proposes the solutions of two problems, namely, the norm-based consistency-restoring problem and the unnormalized consistency-restoring problem. The solution of the first problem consists of two approaches which employ the violation vector or the satisfying restored probability vector to compute new probability for each probabilistic constraint. For the second problem we have proposed three approaches which employ a balanced consistency-restoring operator or equitable consistency-restoring operator or an amerced restoring operator compute new probability for each probabilistic constraint. A set of axioms which represent the relationship between the logical properties and operators have been investigated and discussed. For all proposed approaches, algorithms for restoring the consistency of a PKB as well as its complexity assessment have been worked out. In the next two chapters, we will deal with the probability knowledge integration problems.

CHAPTER 5

Distance-based methods for integrating probabilistic knowledge bases

T HE PURPOSE OF THIS CHAPTER is to handle the probabilistic knowledge inte-gration problem. Firstly, we present the overview of the integration problem and methods for integrating knowledge bases. We then take into account the probabilistic knowledge integration problem. In particular, we propose a general distance-based model for integrating the PKBs and show how to solve the integrating problems. Distance-based integrating operators and their desired properties also are considered to ensure the integrating principles. Finally, fundamental methods such as techniques for discovering probability integrating vector, finding the satisfying probability vector, and the HULL algorithm are used to solve integrating problems. The complexity of algorithms is discussed and proved.

5.1 OVERVIEW OF KNOWLEDGE INTEGRATION METHODS

5.1.1 The Knowledge Integration Problem

It is essential to improve the appropriate methods for building and maintaining the action of knowledge-based systems. In secure systems, the knowledge representa-tion methods are used for the integration of different distributed information sources to make the system more flexible and efficient. When two computer systems must be integrated or the content of their knowledge bases must be exchanged, the infor-mation stored in those knowledge bases must be appropriately modified. Knowledge integration could be understood as a process of creating a consistent knowledge from a set of knowledge bases deriving from different systems [1]. According to Nguyen [1] knowledge integration can be considered on the syntactic and semantic levels. It consists of the knowledge integration problems for the conjunctive, disjunctive, fuzzy conjunctive normal form structure.

The knowledge integration problem [1] on the syntactic level:

(i) The knowledge integration problem for the conjunctive normal form structure is defined as follows:

For a given conflict profile of conjunctions

$$X = \left\{ x_i = (x_i^+, x_i^-) \in Conj\left(L\right), i = 1, ..., n \right\}$$

where L is a finite set representing the positive logical value related to specific real-world events or objects, $Conj(L)$ is the set of all conjunctions with symbols over L, x^+ is a positive element of the conjunction, x^- is a negative element of the conjunction. It is necessary to determine a conjunction $x^* \in Conj$ called a consensus of set X.

(ii) The knowledge integration problem for the disjunctive normal form structure is defined as follows:

For a given set of all clauses

$$X = \left\{ x_i = (x_i^+, x_i^-), i = 1, ..., n \right\}$$

where x^+ is a positive element of a clause, x^- is a negative element of a clause. It is necessary to determine a clause $x^* = (x^{*^+}, x^{*^-})$ called a consensus of set X.

(iii) The knowledge integration problem for the fuzzy conjunctive normal form structure is defined as follows:

For a given set of fuzzy conjunctions

$$X = \left\{ x_i = t_1^{(i)}, v_1^{(i)}) \wedge t_2^{(i)}, v_2^{(i)}) \wedge ... \wedge (t_{k_i}^{(i)}, v_{k_i}^{(i)}), i = 1, ..., n \right\}$$

where $t_i \in L$, $v_i \in [0,1]$; it is necessary to determine a fuzzy conjunction $x^* = (t_1, v_1) \wedge (t_2, v_2) \wedge ... \wedge (t_k, v_k)$ called a consensus of set X.

The knowledge integration problem [1] the semantic level:

(i) The knowledge integration problem for the conjunctive normal form structure is defined as follows:

For a given profile

$$X = \left\{ x_i = (x_i^+, x_i^-) \in Conj\,(A, V)\,, i = 1, ..., n \right\}$$

where (A, V) is real world, $Conj(A, V)$ is a set of all conjunctions of (A, V); it is necessary to determine a conjunction x^* which best represents the given conjunctions.

(ii) The knowledge integration problem for the disjunctive normal form structure is defined as follows:

For a given profile

$$X = \left\{ x_i = (x_i^+, x_i^-) \in Clause\,(A, V)\,, i = 1, ..., n \right\}$$

where $Clause(A, V)$ is a set of all (A, V)-based clauses, it is necessary to determine a clause x^* which best represents the given clauses.

Figure 5.1 shows a general model of a knowledge integration process that takes into account the inconsistency of knowledge and is inherited from the idea in [1]. The input of the model is n PKBs, and the output of the model is a joint knowledge base that best represents the input knowledge bases. The model is performed by the integration algorithm.

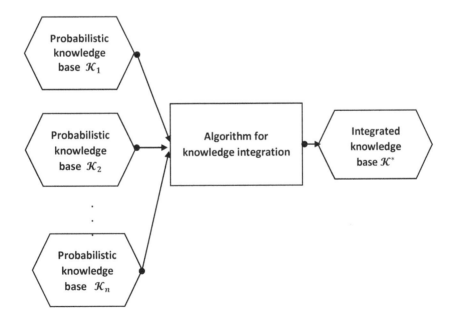

Figure 5.1   The general model for knowledge integration [1].

The probability knowledge integration problem is defined as follows: Given a PKB profile

$$\mathcal{R} = \langle \mathcal{B}, \mathsf{E} \rangle$$

where E is a finite set including n events and $\mathcal{B}$ is a finite multi-set including m PKBs. It is necessary to determine a joint PKB $\mathcal{K}^*$ which best represents the given PKBs. The PKB $\mathcal{K}^*$ is called the knowledge integrated from the PKB profile $\mathcal{R}$.

5.1.2 Methods for Integrating Knowledge Bases

Table 5.1 summarizes methods for integrating knowledge bases according to each type of knowledge base.

Table 5.1 Methods for integrating knowledge bases

Knowledge bases	Methods	Technique	Disadvantages
Logic	Finding the set of the pre-orders	Integrity constraints [14]	PKBs meet the same structure requirement
		Social theory [18]	Low computational efficiency
		Model [14, 17, 78]	Information loss
Probabilistic-logic	Finding collective knowledge	Decision theory [23]	Being only applicable with probability bounds
Probability	Finding Probability value	Median integration[12, 31, 32, 33]	The input knowledge bases is consistent and have a same structure
	Finding joint probability distribution	Iterative methods [24, 25, 26]	Working on probability distributions with the same structure
		Candidacy function [34]	Low computational efficiency
		DDFs [27, 28, 29, 30]	The input knowledge bases must be probability distributions, consistent and structurally similar

Logical knowledge integration method:

The method of integrating logical knowledge bases was proposed in 1985 by Alchourrón et al. [3]. This method is called belief revision *(Belief Revision)*. After that, many other knowledge integration methods have been proposed to solve the problem in specific situations [14, 15, 16, 17, 18, 78].

These methods include model-based knowledge integration [14, 17, 78], knowledge integration with integrity constraints [14] and logic knowledge integration associated with social choice theory [18]. The idea of model-based knowledge integration is to build a family of integration operators (Δ_μ [17], DA^2 [14], Δ_μ^k [78], Δ^{Gmin} [78]) with the desired logical properties. The propositional logic framework, on the other hand, defines model-based integrating operators. All knowledge bases fulfill the same standards of importance, priority, and dependability. Furthermore, knowledge bases with more structure than propositional logic may be required to be integrated.

Possibilistic logic knowledge integration method: In order to solve the problem of integrating propositional logic knowledge bases that do not have the same importance, the weighted approach is one of the most suitable methods. The weighted knowledge bases are usually represented using possibilistic logic or ordinal conditional functions. There are many integrating approaches proposed in possibilistic logic [13, 15, 16, 79, 80, 81]. This idea of these approaches is that using the notion of a possibility distribution, which is a mapping from the set of interpretations Ω to interval [0,1]. S. Benferhat [13] uses integrating operators $\oplus$ to aggregate the possibility distributions associated with them into a new possibility distribution. A family of integrating operators Δ^{PLMIN}, $\Delta^{PLMIN,X}$ [15] is proposed for the knowledge base representing the integrity constraints with respect to a total pre-order. J.F. Du et al. [16] use the lexicographic ordering to propose an integrating operator Δ^{Lex} that maps a formula representing the integrity constraints and a set of possibilistic knowledge bases to a classical knowledge base. Qi et al. [79, 80, 81] employ the maximum (or T-conorm) and the minimum (or T-norm) operators to build a split-combination method. This model consists of two steps. The first step is to split each of them into two subbases according to the upper free degree. The second step is to use T-conorm mode to combine the inconsistent part and T-norm mode to combine the consistent part. However, the weighted information is not useful anymore if a formula can be deduced from the consistent merged knowledge base and do not take into account the inference degree.

Another integration method comes from the natural idea when people resolve conflicts. The solution is that a group of people who have some conflicting opinions that will let them discuss, debate and negotiate with each other to reach consensus [19, 20, 21, 22]. Zhang [20] introduces an axiomatic model of negotiation for qualitative bargaining analysis. *However, the limitation of these methods is that each agent's information can be exposed. That causes the agents to lose their negotiating power.*

Probabilistic-logic knowledge integration method:

Potyka Qi et al. [23] introduce a probabilistic-logic framework for group decision-making. A decision base $\mathcal{D}$ over a set of agents $N = \{1, \ldots, n\}$ is a tuple $(\mathcal{K}, \mathcal{A}, \mathcal{C}, \mathcal{U})$, where $\mathcal{K}$ is the knowledge base of $\mathcal{D}$, $\mathcal{A}$ is a non-empty set of alternatives, $\mathcal{C}$ is a non-empty set of criteria such that $\mathcal{C} \in \mathcal{L}$ is a formula that contains exactly one free variable, $U : \mathcal{C} \to \mathbb{R}^+$ is the utility function that maps each criterion to a non-negative utility vector whose i-th value represents the utility for agent i. Group knowledge of each constraint $(F \,|\, G)$ $\mathcal{B}_G (F \,|\, G) = [l, r]$, where $l \overset{def}{=} \arg\min_{\mathcal{P} \in GM_{\mathcal{K}_0}^{\|\cdot\|}(\mathcal{M})} \frac{\mathcal{P}(FG)}{\mathcal{P}(G)}$ and $r \overset{def}{=} \arg\max_{\mathcal{P} \in GM_{\mathcal{K}_0}^{\|\cdot\|}(\mathcal{M})} \frac{\mathcal{P}(FG)}{\mathcal{P}(G)}$ with $\mathcal{P} > 1$. However, this model only applies to knowledge bases represented by interval probabilistic constraints, i.e. each probabilistic constraint has the form $(F \,|\, G) [l, u]$.

Probability knowledge integration method:

There are other knowledge integration methods applied for probabilistic knowledge bases. Vomlel [24] has proposed an integration model as well as methods applicable to the process of integrating PKBs represented by a set of probability distributions. According to Vomlel, knowledge integration is understood as the process of constructing a common probability distribution function from an input set of consistent probability distributions. This integrating procedure has two critical stages: constructing a JPD from a set of low-dimensional distributions and producing a JPD that embodies domain knowledge. In order to cope with integrating inconsistent PKBs, five iterative methods are proposed such as Convex Conservative Modification (CC), Log-Convex Conservative Modification (LCC), Method of Iterative Arithmetic Averages (AA), Method of Iterative Arithmetic Averages (GA), an instance of Generalized EM-algorithm (GEM). Because GEM algorithm satisfies all the necessary properties, it is used to demonstrate the convergence properties of the iterative method. Let $V = \{1, \ldots, n\}$, $\mathcal{P}$ be a set of low-dimensional distributions, $\varepsilon = \{E_1, \ldots, E_n\}$ be a generating class of input set $\mathcal{P}_\varepsilon$ with $E_i \subseteq V$, and $Q_0 \in \mathcal{P}$ be an initial probability distribution. Let $x \in \mathbb{X}_V = \{\{X_i\}_{i \in V}\}$, $j = ((i-1) \bmod s + 1)$, $P_{E_j}(x^{E_j}) \in \mathcal{P}$. Solving the integrating problem by using IPFP method [24] is a computational process $Q_{(i)}(x)$ such that:

$$Q_{(i)}(x) = \begin{cases} Q_{(i-1)}(x) \dfrac{P_{E_j}(x^{E_j})}{Q_{(i-1)}^{E_j}(x^{E_j})}, & \text{if } x \text{ satisfies } Q_{(i-1)}^{E_j} > 0. \\ 0, & \text{if } x \text{ satisfies } Q_{(i-1)}^{E_j} = 0. \end{cases}$$

Zhang et al. proposed a method, named Smooth algorithm to allow to deal with both consistent and inconsistent constraints [26]. Smooth's convergence, on the other hand, has only been demonstrated empirically through tests, and iterative methods have failed to produce satisfactory results when probability distributions contain zeroes.

In the field of combining information, such as statistics, decision theory and economic science, the problem of combining subjective probabilities has been studied for a long time. To deal with the association problem, methods of maximum entropy have been applied to probabilities represented by random variables [31]. ME principles were used to represent incomplete probabilistic knowledge with the aid of the ME method. [36] However, the probability values do not lie within the interval of the initial minimum and maximum probability values. Kern-Isberner and Rödder [32] introduced a new probabilistic revision operator that is intermediate between revision and updating by using the principles of maximum entropy, specifically, $\mathcal{P}^* = ME(\mathcal{K}_1 \odot \mathcal{K}_2)$.

Daniel's thesis [34] introduces a model to derive a knowledge base from inconsistent propositional knowledge bases represented by candidacy functions, that is, Internal Entropy-based Enference Process Δ_i^E and Paraconsistent Maximum Entropy Inference Process Δ_{ME}^E. This thesis asserts that the proposed approach is the only solution for problems in election theory and can handle the complexity due to the knowledge representation space. *However, as the complex structure and non-linearity of the candidacy function, the computational efficiency is low and the obtained results are approximate.*

Another approach also works with probability distributions but this approach is based on the divergence technique [27, 28, 29]. Adamcik [29] constructs a class of integration operators based on the concept of divergent distances such that they satisfy the integration principles proposed by Konieczny and Pino Perez [17]. This operator is defined by $\Delta(\mathcal{K}_1, \ldots, \mathcal{K}_n) = \{arg\min_{v \in D^L} B(w||v) : w^{(i)} \in V_{\mathcal{K}_i}^L\}$ where B is a Bregman divergence function, and D^L is the set of all probability functions over the finite propositional language L. These operators consist of the linear Euclidean integration operator $\widehat{\Delta}_L^{E2}$, Kullback-Leibler operator $\widehat{\Delta}^{KL}$, Renyi-B operator $\widehat{\Delta}^B$. However, Adamcik has only built a mathematical model that allows potential expert systems to evaluate probability without studying methods to design a system of PKBs. Furthermore, in order to use this mathematical model, the initial assumption is that the probabilistic knowledge of a particular expert must be consistent and obeys the principle of probability. In addition, a general application for knowledge integration and the complexity of the proposed algorithms have not been thoroughly studied and analyzed.

5.2 PROBABILISTIC KNOWLEDGE INTEGRATION

5.2.1 Divergence Functions

Definition 5.1 ([82]) *A function $d : \mathbb{P}(E) \times \mathbb{P}(E) \to \mathbb{R}^*$ is called a divergence distance function (DDF) if it satisfies the following properties:*

1. $d(\vec{x}, \vec{y}) = 0$ iff $\vec{x} = \vec{y}$.

2. $d(\vec{x}, \vec{y}) = d(\vec{y}, \vec{x})$ where $\vec{x}, \vec{y}$ are probability functions.

A comparison study is definitely a difficult assignment for analyzing and evaluating the applicability of DDFs employed in integrating situations. Each of these functions is summarized and grouped into two primary groups based on its properties. The first uses basic mathematical operations like square, absolute calculation, logarithm, and division of probability distribution functions.

Table 5.2 summarizes and presents basic DDFs for building distance-based probabilistic knowledge integration problems.

Table 5.2 Basic DDFs [29, 51, 82, 83, 84, 85, 86, 87]

Functions	Formulas
B-Div	$d^{\mathsf{B}}(\vec{x}, \vec{y}) = 2 - 2 \sum_{i \in I(y)} \sqrt{x_i y_i}$
E-Div	$d^{\mathsf{E}}(\vec{x}, \vec{y}) = \sum_{i \in I(y)} x_i (log x_i - log y_i)^2$
H-Div	$d^{\mathsf{H}}(\vec{x}, \vec{y}) = \frac{1}{\sqrt{2}} \sqrt{\sum_{i \in I(y)} (\sqrt{x_i} - \sqrt{y_i})^2}$
I-Div	$d^{\mathsf{I}}(\vec{x}, \vec{y}) = \sum_{i \in I(y)} \left(\frac{x_i log x_i + y_i log y_i}{2} - \frac{x_i + y_i}{2} log \frac{x_i + y_i}{2} \right)$
J-Div	$d^{\mathsf{J}}(\vec{x}, \vec{y}) = \sum_{i \in I(y)} (x_i - y_i) log \frac{x_i}{y_i}$
K-Div	$d^{\mathsf{K}}(\vec{x}, \vec{y}) = \frac{1}{2} \sum_{i \in I(y)} \frac{(x_i - y_i)^2}{y_i}$
L-Div	$d^{\mathsf{L}}(\vec{x}, \vec{y}) = \sum_{i \in I(y)} x_i log \frac{x_i}{y_i}$
M-Div	$d^{\mathsf{M}}(\vec{x}, \vec{y}) = 1 - \sum_{i \in I(y)} \frac{2 x_i y_i}{x_i + y_i}$
N-Div	$d^{\mathsf{N}}(\vec{x}, \vec{y}) = \sum_{i \in I(y)} x_i log \frac{x_i}{0.5 x_i + 0.5 y_i}$
Q-Div	$d^{\mathsf{Q}}(\vec{x}, \vec{y}) = \sum_{i \in I(y)} (\sqrt{x_i} - \sqrt{y_i})^2$
S-Div	$d^{\mathsf{S}}(\vec{x}, \vec{y}) = \sum_{i=1}^{m} (x_i - y_i)^2$
T-Div	$d^{\mathsf{T}}(\vec{x}, \vec{y}) = \sum_{i \in I(y)} \left(\frac{x_i + y_i}{2} log \frac{x_i + y_i}{2 \sqrt{x_i y_i}} \right)$
V-Div	$d^{\mathsf{V}}(\vec{x}, \vec{y}) = \sum_{i=1}^{m} |x_i - y_i|$

In the first group, there are the following DDFs: The squared euclidean distance is used to create the S-Div function [29], whereas the total variation is used to create the V-Div function [87]. The Kagan divergence is used to create the K-Div function [83]. In terms of square roots, the DDFs include B-Div, Q-Div, H-Div functions. The Bhattacharyya distance is used to create the B-Div function, while the Hellinger distance is used to create the H-Div function. These two probability metrics-based functions are derived from [87]. Jeffreys distance is the foundation of the Q-Div function [84]. In terms of logarithm, the DDFs include the functions, namely E-Div, I-Div, J-Div, L-Div, N-Div, T-Div. The novel directed divergence measure developed is used to construct the N-Div function [85]. The Jensen difference divergence measure is used to construct the I-Div function, whilst the arithmetic and geometric mean divergence measure is used to construct the T-Div function [86]. The Kullback-Leibler divergence is used to define the L-Div function [29], while the symmetric divergence is used to define the J-Div function [84]. The exponential divergence is used to create the E-Div function [83]. In the second group, A-Div, NA-Div, AI-Div, NAI-Div, C-Div, CS-Div, D-Div, and R-Div are among the DDFs of order $alpha$.

Table 5.3 DDFs of order λ [29, 51, 82, 83, 84]

Functions	Formulas	Order
A-Div	$d^{A,\lambda}(\vec{x}, \vec{y}) = \frac{1}{\lambda-1} log(\sum_{i \in I(y)} x_i^{\lambda} y_i^{1-\lambda})$	$\lambda \neq 1$
AI-Div	$d^{AI,\lambda}(\vec{x}, \vec{y}) = \frac{1}{\lambda-1} log(\sum_{i \in I(y)} x_i y_i^{\lambda-1})$	$\lambda \neq 1$
C-Div	$d^{C,\lambda}(\vec{x}, \vec{y}) = \frac{4}{1-\lambda^2} \left(1 - \sum_{i \in I(y)} x_i^{\frac{1-\lambda}{2}} y_i^{\frac{1+\lambda}{2}}\right)$	$\lambda \neq 1$
CS-Div	$d^{CS,\lambda}(\vec{x}, \vec{y}) = \frac{1}{2} \left(1 - \sum_{i \in I(y)} \left((x_i y_i)^{\frac{1}{2}} cos(\lambda log \frac{x_i}{y_i})\right)\right)$	$\frac{-\Pi}{2} \leq \lambda \leq \frac{\Pi}{2}$
D-Div	$d^{D,\lambda}(\vec{x}, \vec{y}) = 1 - \sum_{i \in I(y)} \frac{1}{2^{\lambda}} \frac{x_i y_i}{(x_i^{\lambda} + y_i^{\lambda})^{-\lambda}}$	$\lambda \neq 1$
NA-Div	$d^{NA,\lambda}(\vec{x}, \vec{y}) = \frac{1}{2^{\lambda-1}-1} \left(\sum_{i \in I(y)} x_i^{\lambda} y_i^{1-\lambda} - 1\right)$	$\lambda \neq 1$
NAI-Div	$d^{NAI,\lambda}(\vec{x}, \vec{y}) = \frac{1}{2^{\lambda-1}-1} \left(\sum_{i \in I(y)} x_i y_i^{\lambda-1} - 1\right)$	$\lambda \neq 1$
R-Div	$d^{R,\lambda}(\vec{x}, \vec{y}) = \sum_{i=1}^{m} (x_i^{\lambda} - y_i^{\lambda} - \lambda(x_i - y_i)(y_i)^{\lambda-1})$	$\lambda \in \mathbb{R}_{(1,2]}$

The A-Div and NA-Div are functions defined for $\alpha \neq 1$ using additive directed divergence and non-additive directed divergence of order α, respectively. Similarly, the AI-Div and NAI-Div are functions based on the additive inaccuracy divergence and non-additive inaccuracy divergence of order α, respectively. These functions are first introduced in [82]. The cosine divergence of degree α [84] is used to define the **CS**-Div function, whereas the Chernoff information of order α [83] is used to define the **C**-Div function. The Renyi divergence [29] is used to create the **R**-Div function.

Regarding the link between DDFs, we make the following observations.

- $d^J(\vec{x}, \vec{y}) = d^L(\vec{x}, \vec{y}) + d^L(\vec{y}, \vec{x})$,
- $d^N(\vec{x}, \vec{y}) = d^L(\vec{x}, \frac{\vec{x}+\vec{y}}{2})$, $d^I(\vec{x}, \vec{y}) = \frac{1}{2}(d^L(\vec{x}, \frac{\vec{x}+\vec{y}}{2}) + d^L(\vec{y}, \frac{\vec{x}+\vec{y}}{2}))$,
- $d^T(\vec{x}, \vec{y}) = \frac{1}{2}(d^L(\frac{\vec{x}+\vec{y}}{2}, \vec{x}) + d^L(\frac{\vec{x}+\vec{y}}{2}, \vec{y}))$,
- $d^M(\vec{x}, \vec{y}) \leq \frac{1}{4}(d^L(\vec{x}, \vec{y}) + d^L(\vec{y}, \vec{x}))$,
- $d^V(\vec{x}, \vec{y}) \leq \sqrt{2d^L(\vec{x}, \vec{y})}$,
- $d^Q(\vec{x}, \vec{y}) = d^B(\vec{x}, \vec{y}) \leq 2 - (d^L(\frac{\vec{x}+\vec{y}}{2}, \vec{x}) + d^L(\frac{\vec{x}+\vec{y}}{2}, \vec{y}))$,
- $d^{NAI}(\vec{x}, \vec{y}) = \frac{2^{1-\lambda}d^{AI}-1}{2^{1-\lambda}-1}$

If $\lambda = 0$ then $d^{CS,0}(\vec{x}, \vec{y}) = \frac{1}{4}d^B(\vec{x}, \vec{y})$, $d^{C,0}(\vec{x}, \vec{y}) = 2d^B(\vec{x}, \vec{y})$

If $\lambda = \frac{1}{2}$ then $d^{A,\frac{1}{2}}(\vec{x}, \vec{y}) = -2ln(1 - \frac{1}{2}d^B(\vec{x}, \vec{y}))$, $d^{NA,\frac{1}{2}}(\vec{x}, \vec{y}) = \frac{1}{2-\sqrt{2}}d^B(\vec{x}, \vec{y})$

If $\lambda = 2$ then $d^{A,2}(\vec{x}, \vec{y}) = ln(1+\frac{1}{2}d^K(\vec{x}, \vec{y}))$, $d^{NA,2}(\vec{x}, \vec{y}) = 2d^K(\vec{x}, \vec{y})$, $d^{R,2}(\vec{x}, \vec{y}) = d^S(\vec{x}, \vec{y})$

Consider the challenge of integrating two PKBs $\mathcal{K}_1$ and $\mathcal{K}_2$. Assume that x_i and y_i are the probability function values of the full conjunction Θ_i fulfilling $\mathcal{K}_1$ and $\mathcal{K}_2$. Consider DDF d^L, where $- \sum_{i \in I(y)} x_i log y_i$ might be read as the probability function values that one expects to receive. As a result, $\sum_{i \in I(y)} x_i log x_i - \sum_{i \in I(y)} x_i log y_i = \sum_{i \in I(y)} x_i log \frac{x_i}{y_i}$ can be the joint probability function values of $\mathcal{K}_1$ and $\mathcal{K}_2$.

Some of the DDFs' features used to characterize these functions have been adopted from [82] and reworked to fit a probabilistic framework.

Definition 5.2 *[51, 82] Given function $d : \mathbb{P}^{\hbar_E} \times \mathbb{P}^{\hbar_E} \to \mathbb{R}^+$; $\vec{x}, \vec{y} \in \mathbb{P}^{\hbar_E}$ where $\vec{x} = (x_1, \ldots, x_m)$ and $\vec{y} = (y_1, \ldots, y_m)$; π is an arbitrary permutation of $1, 2, \ldots, m$. We should have the following properties:*

(NON)- *Non-negativity $d(\vec{x}, \vec{y}) \geq 0$, $d(\vec{x}, \vec{y}) = 0$ iff $\vec{x} = \vec{y}$*

*The property **NON** assures that DDFs is always non-negative.*

(EXP)- *Expansibility $d((\vec{x}, 0), (\vec{y}, 0)) = d(\vec{x}, \vec{y})$.*

It states that DDFs have no modification if a probability having zero value is inserted.

(SYM)- *Symmetry $d(\vec{x}, \vec{y}) = d((x_{\pi(1)}, \ldots, x_{\pi(m)}), (y_{\pi(1)}, \ldots, y_{\pi(m)}))$.*

This property ensures that DDFs are independent of the order of labeled probabilities.

(ADD)- Additivity Let $\vec{z} = (z_1, \ldots, z_n)$, $\vec{t} = (t_1, \ldots, t_n)$ *such that* $\sum_{i=1}^{m} x_i = \sum_{j=1}^{n} z_j = \sum_{i=1}^{m} y_i = \sum_{j=1}^{n} t_i$. *If* $\vec{v} = (x_1 z_1, \ldots, x_1 z_n, \ldots, x_m z_1, \ldots, x_m z_n)$, $\vec{w} = (y_1 t_1, \ldots, y_1 t_n, \ldots, y_m t_1, \ldots, y_m t_n)$ *then* $d(\vec{v}, \vec{w}) = d(\vec{x}, \vec{y}) + d(\vec{z}, \vec{t})$

According to the property **ADD**, *it can be associated with probabilities relating to mutually independent events.*

(SADD)- Strong Additivity Let $\vec{x} = (x_1, \ldots, x_m)$, $\vec{y} = (y_1, \ldots, y_m)$. *For* $\sum_{i=1}^{n} \sum_{j=1}^{m} x_{ij} = 1$, $x_j = \sum_{i=1}^{n} x_{ij} \geq 0$ *and* $\sum_{i=1}^{n} \sum_{j=1}^{m} y_{ij} = 1$, $y_j = \sum_{i=1}^{n} y_{ij} \geq 0$. *If* $\vec{v} = (x_{11}, \ldots, x_{nm})$, $\vec{w} = (y_{11}, \ldots, y_{nm})$, $z_j = (\frac{x_{1j}}{x_j}, \ldots, \frac{x_{nj}}{x_j})$ *and* $t_j = (\frac{y_{1j}}{y_j}, \ldots, \frac{y_{nj}}{y_j})$ *then* $d(\vec{v}, \vec{w}) = d(\vec{x}, \vec{y}) + \sum_{j=1}^{m} x_j d(z_j, t_j)$.

(COT)- Continuity $d(\vec{x}, \vec{y})$ *is continuous.*

Continuity states that if there exist a slight change of the probabilities, DDFs could be altered somewhat.

(NADD)- Non-additivity Let $\vec{x} = (x_1, \ldots, x_m)$, $\vec{z} = (z_1, \ldots, z_n)$, $\vec{y} = (y_1, \ldots, y_m)$, $\vec{t} = (t_1, \ldots, t_n)$ *such that* $\sum_{i=1}^{m} x_i = \sum_{j=1}^{n} z_j = \sum_{i=1}^{m} y_i = \sum_{j=1}^{n} t_i$. *If* $\vec{v} = (x_1 z_1, \ldots, x_1 z_n, \ldots, x_m z_1, \ldots, x_m z_n)$, $\vec{w} = (y_1 t_1, \ldots, y_1 t_n, \ldots, y_m t_1, \ldots, y_m t_n)$ *then* $d(\vec{v}, \vec{w}) = d(\vec{x}, \vec{y}) + d(\vec{z}, \vec{t}) + (2^{\lambda-1} - 1) d(\vec{x}, \vec{y}) d(z, t)$

(SNADD)- Strong Non-additivity Let $\vec{x} = (x_1, \ldots, x_m)$, $\vec{y} = (y_1, \ldots, y_m)$ *such that* $\sum_{i=1}^{m} x_i = \sum_{i=1}^{m} y_i = 1$. *Let* $z_i = (x_{1i}, \ldots, x_{ni})$, $t_i = (y_{1i}, \ldots, y_{ni})$ *such that* $\sum_{j=1}^{n} x_{ji} = \sum_{j=1}^{n} y_{ji} = 1 \; \forall i = 1, \ldots, m$. *If* $\vec{v} = (x_1 x_{11}, \ldots, x_1 x_{n1}, \ldots, x_m x_{1m}, \ldots, x_m x_{nm})$ *and* $\vec{w} = (y_1 y_{11}, \ldots, y_1 y_{n1}, \ldots, y_m y_{1m}, \ldots, y_m y_{nm})$ *then* $d(\vec{v}, \vec{w}) = d(\vec{x}, \vec{y}) + \sum_{i=1}^{m} x_i y_i^{1-\lambda} d(\vec{z}, \vec{t})$.

The logical relationship among desirable characteristics and DFFs is summarized in Table 5.4.

Table 5.4 The logical relationship among desirable characteristics and DFFs [51]

DDF	NON	EXP	SYM	ADD	SADD	COT	NADD	SNADD
A-Div	x	x	x	x	x	x	x	x
AI-Div	x	x	x	x	x	x	x	x
B-Div	x	x	x	x	x	x	-	-
C-Div	x	x	x	x	x	x	-	-
CS-Div	x	x	-	-	-	x	-	-
D-Div	x	x	x	x	x	x	-	-
E-Div	x	x	x	x	x	x	-	-
H-Div	x	x	x	x	x	x	-	-
I-Div	x	x	x	x	x	x	-	-
J-Div	x	x	x	x	x	x	-	-
K-Div	x	x	x	x	x	x	-	-
L-Div	x	x	x	x	x	x	-	-
N-Div	x	x	x	x	x	x	-	-
NA-Div	x	x	x	x	x	x	x	x
NAI-Div	x	x	x	x	x	x	x	x
M-Div	x	x	x	x	x	x	-	-
Q-Div	x	x	x	x	x	x	-	-
R-Div	x	x	x	x	x	x	-	-
S-Div	x	x	x	x	x	x	-	-
T-Div	x	x	x	x	x	x	-	-
V-Div	x	x	x	x	x	x	-	-

5.2.2 Distance-based Model for Integrating Probabilistic Knowledge Bases

The distance-based probabilistic knowledge integration problem is defined as follows:

(1) **Input:** A PKB profile $\mathcal{R} = \langle \mathcal{B}, \mathsf{E} \rangle$.

(2) **Output:** $\mathcal{K} = \{(c_1)[\rho_1], \ldots, (c_m)[\rho_m]\}$ such that $\mathcal{K} \not\models \bot$.

(3) **Scope of problem:** The PKBs are represented by probabilistic constraints.

(4) **The integration process:**

- Step 1: For each $\mathcal{K}_i \in \mathcal{B}$, an inconsistency measure $\mathcal{I}(\mathcal{K}_i)$ is computed by employing Theorem 3.10, Theorem 3.11, Theorem 3.12, and Theorem 3.14.

- Step 2: If $\mathcal{I}(\mathcal{K}_i) \neq 0$, inconsistencies of $\mathcal{K}_i$ are solved by employing Theorem 4.6, Theorem 4.7, and Definition 4.5.

- Step 3: For each $\mathcal{K}_i \in \mathcal{B}$, the satisfying restored probability vector of $\mathcal{K}_i$ is found by employing Theorem 5.2, Theorem 5.3, Theorem 5.4 and Theorem 5.5.

- Step 4: The PKBs in $\mathcal{B}$ is integrated by employing Theorem 5.8 and Theorem 5.9, or Theorem 5.12.

- Step 5: New probability for each probabilistic constraint in the resulting PKB is computed by by employing Theorem 2.1.

A general distance-based model for integrating the PKBs is presented in Figure 5.2.

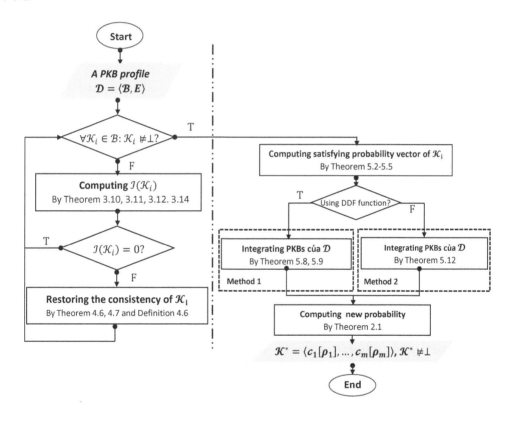

Figure 5.2 The distance-based model for integrating probabilistic knowledge bases [51, 88].

From the model, it is easy to see that both integration methods are based on the satisfying restored probability vector of the input PKB. The first method solves the DDF-based non-linear optimization problem to find the probability integrating vector of a PKB profile. The second one employs the a vector random generation-based formula that guarantees the probability rules to find this integrating vector.

The next section of this chapter will present the theoretical basis for building the algorithms implemented in Step 3, Step 4 and Step 5 of the integration process.

5.2.3 Desired Properties of Distance-based Probabilistic Integrating Operator

The desirable properties which are used to characterize the probabilistic integrating operators is presented in Definition 5.3.

Definition 5.3 ([1, 29, 51, 88]) *Let* $\mathcal{R}_1 = \langle \mathcal{B}, \mathsf{E} \rangle$ *and* $\mathcal{R}_2 = \langle \mathcal{C}, \mathsf{E} \rangle$ *be PKB profiles where* $\mathcal{B} = \{\mathcal{K}_1, \ldots, \mathcal{K}_n\}$ *and* $\mathcal{C} = \{\mathcal{K}'_1, \ldots, \mathcal{K}'_m\}$. *Let* $\mathcal{K}$ *be a set of the finite multisets of PKBs. Function* $\Gamma : \mathbb{B} \to 2^{\mathbb{P}^{\hbar_E}}$ *is called a probabilistic integrating operator iff the following properties are satisfied:*

(CP)-Consistency Principle *If* $\cap_{i=1}^{n} \mho(\mathcal{K}_i) \neq \emptyset$ *then* $\Gamma(\mathcal{B}) \subseteq \cap_{i=1}^{n} \mho(\mathcal{K}_i)$.

> *Consistency principle means that there is the consistent input PKB profile, the integrating result could be the intersection subset of the set of all probability functions fulfilling all PKBs.*

(SCP)-Strong Consistency Principle
> *If* $\cap_{i=1}^{n} \mho(\mathcal{K}_i) \neq \emptyset$ *then* $\Gamma(\mathcal{B}) = \cap_{i=1}^{n} \mho(\mathcal{K}_i)$.

> *It implies that there is the consistent input PKB profile, the integrating result could be the intersection of the set of all probability functions fulfilling all PKBs.*

(EIP)-Empty Invariance Principle
> $\forall \mathcal{K} = \emptyset: \Gamma(\mathcal{B} \cup \mathcal{K}) = \Gamma(\mathcal{B})$

> *This property* **EIP** *says that if an empty PKB is appended to a PKB profile, the result of the integration process are not affected.*

(EP)-Equivalence Principle *If there exists a permutation* μ *over* $\{1, \ldots, n\}$ *such that* $\mho(\mathcal{K}_i) = \mho\left(\mathcal{K}'_{\mu(i)}\right)$ *then* $\Gamma(\mathcal{B}) = \Gamma(\mathcal{C})$.

> *Equivalence principle means that the occurrence order of PKBs does not affect the result of the integration process.*

(PIP)-Positive Invariance Principle
> *If* $\cap_{i=1}^{\hbar_c} \mho(\mathcal{K}'_i) \neq \emptyset$, $\Gamma(\mathcal{B}) \neq \emptyset$ *and* $\Gamma(\mathcal{B}) \subseteq \cap_{i=1}^{\hbar_c} \mho(\mathcal{K}'_i)$ *then* $\Gamma(\mathcal{B} \cup \mathcal{C}) = \Gamma(\mathcal{B})$.

> *Positive invariance principle means that if the PKBs in* $\mathcal{C}$ *are appended to* $\mathcal{B}$ *and the intersection of the probability functions in* $\mathcal{C}$ *contains the integrating result of* $\mathcal{B}$ *then the integrating outcome of the PKBs in* $\mathcal{B}$ *will not be affected.*

(AP)-Agreement Principle

If $\Gamma(\mathcal{B}) \cap \Gamma(\mathcal{C}) \neq \emptyset$ then $\Gamma(\mathcal{B}) \cap \Gamma(\mathcal{C}) = \Gamma(\mathcal{B} \cup \mathcal{C})$

It implies that if there is no common probabilistic constraint of the integrating result of two sets of the PKBs, then the intersection of those two results is similar to the integrating outcome of all the PKBs contained in the two sets of the PKBs.

(DP)-Disagreement Principle

If $\cap_{i=1}^{hc} \mho(K_i') \neq \emptyset$ and $\mho(\Gamma(\mathcal{B})) \neq \mho(\Gamma(\mathcal{C}))$ then $\Gamma(\mathcal{B}) \cap \Gamma(\mathcal{C}) = \emptyset$.

Disagreement principle means that if the integrating result of $\mathcal{B}$ is inconsistent with the integrating result of $\mathcal{C}$ and the PKBs in $\mathcal{C}$ are collectively consistent, then the intersection of the integrating results is empty.

(SDP)-Strong Disagreement Principle

If $\Gamma(\mathcal{B}) \cap \Gamma(\mathcal{C}) \neq \emptyset$ then $\Gamma(\mathcal{B} \cup \mathcal{C}) \cap \Gamma(\mathcal{B}) = \emptyset$.

This property implies that the intersection of the integrating results should be empty.

The following theorem ensures that after the integration process, the probability value of the probabilistic constraints in the resulting PKB should be either a mean value or a value within the interval of the upper bound (the smallest probability value of probabilistic constraints) and the lower bound (the largest probability value of probabilistic constraints). As a result, the probability values in the final PKB represent both agreement and majority viewpoints, because the more experts agree on a certain probability value, the closer the resulting probability will be to that value. Therefore, it is necessary to use a Δ parameter as small as possible to adjust so that the output from the integration process satisfies the above median principle.

Theorem 5.1 ([51, 88]) *Let $\mathcal{R} = \langle \mathcal{B}, \mathsf{E} \rangle$ be a PKB profile and $\mathcal{B} = \{\mathcal{K}_1, \ldots, \mathcal{K}_n\}$. Let $\mathcal{K}$ be the resulting PKB after the integration process. For each $(c)[\rho] \in \mathcal{K}$:*

$$\min_{1 \leq i \leq n} \{\rho_j | (c)[\rho_j] \in \mathcal{K}_i\} - \Delta \leq \rho \leq \max_{1 \leq i \leq n} \{\rho_j | (c)[\rho_j] \in \mathcal{K}_i\} + \Delta \qquad (5.1)$$

where Δ is a standard deviation as small as possible.

Proof 5.1 *Consider $(F|G)[\rho] \in \mathcal{K}$. Assume that $(F|G)[\rho_i] \in \mathcal{K}_i \; \forall \mathcal{K}_i \in \mathcal{B}$. We need to prove that $\min\{\rho_1, \ldots, \rho_n\} - \Delta \leq \rho \leq \max\{\rho_1, \ldots, \rho_n\} + \Delta$ with Δ is a standard* *deviation as small as possible. Let $\underline{m} = \min\{\rho_1, \ldots, \rho_n\}$ and $\overline{m} = \max\{\rho_1, \ldots, \rho_n\}$. Since $\rho_i \in \mathbb{R}_{[0,1]}$, it follows that $\rho \in \mathbb{R}_{[0,1]}$.*

Assume that $\overline{m} \leq \rho$. It follows that $\rho \geq \rho_i \ \forall \ 1 \leq i \leq n$. Hence, $\rho > \rho_i + \Delta$. It is easy see that ρ is originated from ρ_i by using DDFs and probability rules (by Theorem 5.8 and Theorem 5.9) or Formula 5.6 (by Theorem 5.12). Then, $\exists \rho_i$ such that $\rho_i \leq \rho$, which contradicts the hypothesis that $\rho \leq \rho_i \ \forall \ 1 \geq i \leq n$. Therefore, $\rho \leq \overline{m}$.

Similarly, $\rho \leq \underline{m}$. Therefore, $\min\{\rho_1, \ldots, \rho_n\} - \Delta \leq \rho \leq \max\{\rho_1, \ldots, \rho_n\} + \Delta$ □

5.2.4 Finding the Satisfying Probability Vector

This section presents the concept of the satisfying probability vector of a PKB and its related properties. This vector is the input of the integration process. The following definition represents that vector.

Definition 5.4 ([51, 88]) *Given a PKB profile $\mathcal{R} = \langle \mathcal{B}, \mathsf{E} \rangle$, a PKB $\mathcal{K} \in \mathcal{B}$ and $\mathcal{K}$ is consistent. A satisfying probability vector of $\mathcal{K}$ over E is $\vec{\omega}_{\mathcal{K}}^{\mathsf{E}} = (\mathcal{P}_{\mathcal{K}}(\Theta_1), \ldots, \mathcal{P}_{\mathcal{K}}(\Theta_{\hbar_E}))$ such that $\sum_{i=1}^{\hbar_E} \mathcal{P}_{\mathcal{K}}(\Theta_i) = 1$ and $\mathcal{K} = \varphi_{\mathcal{K}}^{\mathsf{E}}(\vec{\omega}_{\mathcal{K}}^{\mathsf{E}})$.*

Intuitively, Definition 5.4 shows that it is possible to use the probability rules from the Theorem 2.1 to find a consistent PKB corresponding to its satisfying probability vector.

Example 5.1 *Consider $\mathcal{K}_2$ in Example 2.3. Consider $\mathcal{P}_{\mathcal{K}_2}(HTD) = 0.25$, $\mathcal{P}_{\mathcal{K}_2}(HT\bar{D}) = 0.3$, $\mathcal{P}_{\mathcal{K}_2}(H\bar{T}D) = 0.15$, $\mathcal{P}_{\mathcal{K}_2}(H\bar{T}\bar{D}) = 0$, $\mathcal{P}_{\mathcal{K}_2}(\bar{H}TD) = 0.05$, $\mathcal{P}(\bar{H}T\bar{D}) = 0$, $\mathcal{P}_{\mathcal{K}_2}(\bar{H}\bar{T}D) = 0.05$, $\mathcal{P}_{\mathcal{K}_2}(\bar{H}\bar{T}\bar{D}) = 0.2$. We have,*
$\vec{\omega}_{\mathcal{K}_2}^{\mathsf{E}} = (0.25, 0.30, 0.15, 0, 0.05, 0, 0.05, 0.20)$.

Then, $\mathcal{P}_{\mathcal{K}_2}(HTD) + \mathcal{P}_{\mathcal{K}_2}(HT\bar{D}) + \mathcal{P}_{\mathcal{K}_2}(H\bar{T}D) + \mathcal{P}_{\mathcal{K}_2}(H\bar{T}\bar{D}) + \mathcal{P}_{\mathcal{K}_2}(\bar{H}TD) + \mathcal{P}_{\mathcal{K}_2}(\bar{H}T\bar{D}) + \mathcal{P}_{\mathcal{K}_2}(\bar{H}\bar{T}D) + \mathcal{P}_{\mathcal{K}_2}(\bar{H}\bar{T}\bar{D}) = 1$.

By rule (P0) in Definition 2.1, $\mathcal{P}(H) = \mathcal{P}_{\mathcal{K}_2}(HTD) + \mathcal{P}_{\mathcal{K}_2}(HT\bar{D}) + \mathcal{P}_{\mathcal{K}_2}(H\bar{T}D) + \mathcal{P}_{\mathcal{K}_2}(H\bar{T}\bar{D}) = 0.25 + 0.30 + 0.15 + 0 = 0.7$, $\mathcal{P}(T) = 0.6$, $\mathcal{P}(D) = 0.5$.

By rule (P1) in Definition 2.1, $\mathcal{P}(TH) = \mathcal{P}_{\mathcal{K}_2}(HTD) + \mathcal{P}_{\mathcal{K}_2}(HT\bar{D}) = 0.55$, $\mathcal{P}(HD) = 0.4$,

By rule (P2) in Definition 2.1, $\mathcal{P}(T|H) = \frac{0.55}{0.7} = 0.78$, $\mathcal{P}(H|D) = 0.8$.

Therefore, $\mathcal{K}_2$ is drawn from $\vec{\omega}_{\mathcal{K}_2}^{\mathsf{E}}$.

Theorem 5.2 states that the probability vector with respect to p-norm ($p > 1$ and $p \neq \infty$) of a consistent PKB is its satisfying probability vector.

Theorem 5.2 ([51]) *Given a PKB profile $\mathcal{R} = \langle \mathcal{B}, \mathsf{E} \rangle$ and a PKB $\mathcal{K} \in \mathcal{B}$. If $\mathcal{K}$ is consistent then the probability vector $\vec{\omega}_{\mathcal{K}}^{\mathsf{E},p}$ with respect to p-norm (p > 1 and $p \neq \infty$) of $\mathcal{K}$ over E is the satisfying probability vector of $\mathcal{K}$.*

Proof 5.2 *By Theorem 4.2 there exists $\vec{\omega}_{\mathcal{K}}^{\mathsf{E},p} = (\mathcal{P}(\Theta_1), \dots, \mathcal{P}(\Theta_{\hbar_{\mathsf{E}}}))$. As $\mathcal{K} \not\models \bot$, by applying the definition of the probabilistic inference function (Definition 4.1), we have $\varphi_{\mathcal{K}}^{\mathsf{E}}(\vec{\omega}_{\mathcal{K}}^{\mathsf{E},p}) = \mathcal{K}$. Therefore, by Definition 5.4, $\vec{\omega}_{\mathcal{K}}^{\mathsf{E},p}$ is the satisfying probability vector with respect to p-norm of $\mathcal{K} \in \mathcal{B}$ over E.* □

Theorem 5.3 states that the probability vector with respect to 1-norm of a consistent PKB is its satisfying probability vector with respect to 1-norm.

Theorem 5.3 ([51]) *Given a PKB profile $\mathcal{R} = \langle \mathcal{B}, \mathsf{E} \rangle$ and a PKB $\mathcal{K} \in \mathcal{B}$. If $\mathcal{K}$ is consistent then the probability vector $\vec{\omega}_{\mathcal{K}}^{\mathsf{E},1}$ with respect to 1-norm of $\mathcal{K}$ over E is the satisfying probability vector.*

Proof 5.3 *By Theorem 4.3 there exists $\vec{\omega}_{\mathcal{K}}^{\mathsf{E},1} = (\mathcal{P}(\Theta_1), \dots, \mathcal{P}(\Theta_{\hbar_{\mathsf{E}}}))$. As $\mathcal{K} \not\models \bot$, by applying the definition of the probabilistic inference function (Definition 4.1), we have $\varphi_{\mathcal{K}}^{\mathsf{E}}(\vec{\omega}_{\mathcal{K}}^{\mathsf{E},1}) = \mathcal{K}$. Therefore, by Definition 5.4, $\vec{\omega}_{\mathcal{K}}^{\mathsf{E},1}$ is the satisfying probability vector with respect to 1-norm of $\mathcal{K} \in \mathcal{B}$ over E.* □

Theorem 5.4 states that the probability vector with respect to ∞-norm of a consistent PKB is its satisfying probability vector with respect to ∞-norm.

Theorem 5.4 ([51]) *Given a PKB profile $\mathcal{R} = \langle \mathcal{B}, \mathsf{E} \rangle$ and a PKB $\mathcal{K} \in \mathcal{B}$. If $\mathcal{K}$ is consistent then the probability vector $\vec{\omega}_{\mathcal{K}}^{\mathsf{E},\infty}$ with respect to ∞-norm of $\mathcal{K}$ over E is the satisfying probability vector.*

Proof 5.4 *By Theorem 4.4 there exists $\vec{\omega}_{\mathcal{K}}^{\mathsf{E},\infty} = (\mathcal{P}(\Theta_1), \dots, \mathcal{P}(\Theta_{\hbar_{\mathsf{E}}}))$. Vì $\mathcal{K} \not\models \bot$, using the definition of the probabilistic inference function (Definition 4.1), we have $\varphi_{\mathcal{K}}^{\mathsf{E}}(\vec{\omega}_{\mathcal{K}}^{\mathsf{E},\infty}) = \mathcal{K}$. Therefore, by Definition 5.4, $\vec{\omega}_{\mathcal{K}}^{\mathsf{E},\infty}$ is the satisfying probability vector with respect to ∞-norm of $\mathcal{K} \in \mathcal{B}$ over E.* □

Theorem 5.5 states that the unnormalized probability vector of a consistent PKB is its unnormalized satisfying probability vector.

Theorem 5.5 ([51]) *Given a PKB profile $\mathcal{R} = \langle \mathcal{B}, \mathsf{E} \rangle$ and a PKB $\mathcal{K} \in \mathcal{B}$. If $\mathcal{K}$ is consistent then the unnormalized probability vector $\vec{\omega}_{\mathcal{K}}^{\mathsf{E},u}$ of $\mathcal{K}$ over E is the unnormalized satisfying probability vector.*

Proof 5.5 *By Theorem 4.9 there exists $\vec{\omega}_{\mathcal{K}}^{E,u} = (\mathcal{P}(\Theta_1), \ldots, \mathcal{P}(\Theta_{\hbar_E}))$. As $\mathcal{K} \not\models \bot$, by applying the definition of the probabilistic inference function (Definition 4.1), we have $\varphi_{\mathcal{K}}^{E}(\vec{\omega}_{\mathcal{K}}^{E,u}) = \mathcal{K}$. Therefore, by Definition 5.4, $\vec{\omega}_{\mathcal{K}}^{E,u}$ is the unnormalized satisfying probability vector of $\mathcal{K} \in \mathcal{B}$ over E.* □

Theorem 5.6 implies that if there exists two satisfying probability vectors of a consistent PKB then a PKB obtained from a vector are equivalent to other obtaining from the remaining vector.

Theorem 5.6 ([51]) *Let $\mathcal{R} = \langle \mathcal{B}, E \rangle$ a PKB profile, $\mathcal{K} \in \mathcal{B}$ and $\mathcal{K} \not\models \bot$. If there exist any two satisfying probability vectors $\vec{\omega}_{\mathcal{K}}^{E}$ and $\vec{\omega}'_{\mathcal{K}}^{E}$ then $\varphi_{\mathcal{K}}^{E}(\vec{\omega}_{\mathcal{K}}^{E}) \cong \varphi_{\mathcal{K}}^{E}(\vec{\omega}'_{\mathcal{K}}^{E})$.*

Proof 5.6 *By Theorem 5.2, Theorem 5.3, and Theorem 5.4, a satisfying probability vector of consistent PKB $\mathcal{K}$ corresponds to an optimal solution part of optimization problems. Therefore, $|\varphi_{\mathcal{K}}^{E}(\vec{\omega}_{\mathcal{K}}^{E})| = |\varphi_{\mathcal{K}}^{E}(\vec{\omega}'_{\mathcal{K}}^{E})|$. By Definition 3.16, $\exists \vec{\vartheta} \in \mathbb{R}_{[0,1]} :$ $\varphi_{\mathcal{K}}^{E}(\vec{\omega}_{\mathcal{K}}^{E}) = \partial_{\varphi_{\mathcal{K}}^{E}(\vec{\omega}'_{\mathcal{K}}^{E})}(\vec{\vartheta})$. Therefore, by Definition 3.17 about the qualitatively equivalent of two PKBs $\varphi_{\mathcal{K}}^{E}(\vec{\omega}_{\mathcal{K}}^{E})$ and $\varphi_{\mathcal{K}}^{E}(\vec{\omega}'_{\mathcal{K}}^{E})$, we have $\varphi_{\mathcal{K}}^{E}(\vec{\omega}_{\mathcal{K}}^{E}) \cong \varphi_{\mathcal{K}}^{E}(\vec{\omega}'_{\mathcal{K}}^{E})$.* □

Theorem 5.7 shows that if there exists a satisfying probability vector of a consistent PKB and if there exists a satisfying restored probability vector of this PKB then a PKB obtained from the satisfying probability vector is equivalent to a PKB obtained from the satisfying restored probability vector.

Theorem 5.7 ([51]) *Let $\mathcal{R} = \langle \mathcal{B}, E \rangle$ a PKB profile and $\mathcal{K} \in \mathcal{B}$. If there exists a satisfying probability vector $\vec{\omega}_{\mathcal{K}}^{E}$ and a satisfying restored probability vector $\vec{\sigma}_{\mathcal{K}}^{E}$ then there exists $\mathcal{K}'$ such that $\mathcal{K}' \not\models \bot$ and $\varphi_{\mathcal{K}'}(\vec{\sigma}_{\mathcal{K}}^{E}) \cong \varphi_{\mathcal{K}}^{E}(\vec{\omega}_{\mathcal{K}}^{E})$.*

Proof 5.7 *By Definition 5.4, if there exists $\vec{\omega}_{\mathcal{K}}^{E}$ then $\varphi_{\mathcal{K}}^{E}(\vec{\omega}_{\mathcal{K}}^{E}) = \{c_1[\vartheta_1], \ldots, c_{\bar{b}_{\mathcal{K}}}[\vartheta_{\bar{b}_{\mathcal{K}}}]\}$. By Definition 4.3, if there exists $\vec{\sigma}_{\mathcal{K}}^{E}$ then $\varphi_{\mathcal{K}'}^{E}(\vec{\sigma}_{\mathcal{K}}^{E}) = \{c_1[\vartheta'_1], \ldots, c_{\bar{b}_{\mathcal{K}}}[\vartheta'_{\bar{b}_{\mathcal{K}}}]\}$. Hence, $|\varphi_{\mathcal{K}'}^{E}(\vec{\sigma}_{\mathcal{K}}^{E})| = |\varphi_{\mathcal{K}}^{E}(\vec{\omega}_{\mathcal{K}}^{E})|$. By Definition 3.16, $\exists \vec{\vartheta} \in \mathbb{R}_{[0,1]} : \varphi_{\mathcal{K}'}^{E}(\vec{\sigma}_{\mathcal{K}}^{E}) = \partial_{\varphi_{\mathcal{K}}^{E}(\vec{\omega}_{\mathcal{K}}^{E})}(\vec{\vartheta})$. Therefore, by Definition 3.17 about the qualitatively equivalent of two PKBs $\varphi_{\mathcal{K}'}^{E}(\vec{\sigma}_{\mathcal{K}}^{E})$ and $\varphi_{\mathcal{K}}^{E}(\vec{\omega}_{\mathcal{K}}^{E})$, we have $\varphi_{\mathcal{K}'}^{E}(\vec{\sigma}_{\mathcal{K}}^{E}) \cong \varphi_{\mathcal{K}}^{E}(\vec{\omega}_{\mathcal{K}}^{E})$.* □

Example 5.2 *Let's continue from Example 4.4. From Table 4.5, by Theorem 5.3, the satisfying probability vector $\vec{\omega}_{\mathcal{K}}^{E,1}$ with respect to 1-norm of $\mathcal{K}_1$ over E corresponds with $\vec{\omega}^*$ of the following linear optimization problem:*

$$\arg \min_{(\vec{\omega}, \vec{\lambda}) \in \mathbb{R}^{8+5}} (\lambda_1 + \lambda_2 + \lambda_3 + \lambda_4 + \lambda_5) \tag{5.2}$$

subject to: (3.29), (3.31-3.34), (3.36-3.41), (3.30) is replaced by $0.5\omega_1 + 0.5\omega_2 -$ $0.45\omega_3 - 0.45\omega_4 + 0.5\omega_5 + 0.5\omega_6 - 0.45\omega_7 - 0.45\omega_8 - \lambda_2 \leq 0$, (3.35) is replaced by $0.5\omega_1 + 0.5\omega_2 - 0.45\omega_3 - 0.45\omega_4 + 0.5\omega_5 + 0.5\omega_6 - 0.45\omega_7 - 0.45\omega_8 + \lambda_2 \geq 0$.

The constraints (3.30) and (3.35) are changed because the constraint $(T)[0.3]$ in $\mathcal{K}_1$ is changed to the constraint $(T)[0.45]$ after the consistency-restoring process of $\mathcal{K}_1$.

The optimal solution of problem (5.2) is

$(\vec{\omega}^*, \vec{\lambda}^*) = (0, 0.35, 0.29, 0.06, 0.10, 0, 0.06, 0.14, 0, 0, 0, 0, 0)$,

where

$\vec{\omega}^* = (0, 0.35, 0.29, 0.06, 0.10, 0, 0.06, 0.14)$ and

$\vec{\lambda}^* = (0, 0, 0, 0, 0)$.

Therefore, the satisfying probability vector $\vec{\omega}_{\mathcal{K}_1}^{E,1}$ with respect to 1-norm of $\mathcal{K}_1$ over E is $\vec{\omega}^* = (0, 0.35, 0.29, 0.06, 0.10, 0, 0.06, 0.14)$.

It follows that $\vec{\lambda}^* = (0, 0, 0, 0, 0)$, that is, $\mathcal{I}_E^1(\eta^1(\mathcal{K}_1)) = 0$ which does not contradict the assumption that $\eta^1(\mathcal{K}_1)$ is consistent.

The satisfying probability vector with respect to 1-norm, 2-norm, ∞-norm, and the unnormalized form of $\mathcal{K}_1, \mathcal{K}_2, \mathcal{K}_3, \mathcal{K}_4$, and $\mathcal{K}_5$ over E are given by Table 5.5.

Table 5.5 The satisfying probability vectors of $\mathcal{K}_1, \mathcal{K}_2, \mathcal{K}_3, \mathcal{K}_4, \mathcal{K}_5$ [51]

Vector	HTD	$HT\bar{D}$	$H\bar{T}D$	$H\bar{T}\bar{D}$	$\bar{H}TD$	$\bar{H}T\bar{D}$	$\bar{H}\bar{T}D$	$\bar{H}\bar{T}\bar{D}$	Sum
$\vec{\omega}_{\mathcal{K}_1}^{E,1}$	0.00	0.35	0.29	0.06	0.10	0.00	0.06	0.14	1.00
$\vec{\omega}_{\mathcal{K}_2}^{E,1}$	0.25	0.30	0.15	0.00	0.05	0.00	0.05	0.20	1.00
$\vec{\omega}_{\mathcal{K}_3}^{E,1}$	0.25	0.13	0.19	0.00	0.31	0.00	0.00	0.12	1.00
$\vec{\omega}_{\mathcal{K}_4}^{E,1}$	0.54	0.00	0.04	0.12	0.06	0.00	0.00	0.24	1.00
$\vec{\omega}_{\mathcal{K}_5}^{E,1}$	0.00	0.34	0.35	0.11	0.05	0.00	0.10	0.05	1.00
$\vec{\omega}_{\mathcal{K}_1}^{E,2}$	0.14	0.21	0.14	0.21	0.05	0.04	0.11	0.10	1.00
$\vec{\omega}_{\mathcal{K}_2}^{E,2}$	0.30	0.25	0.10	0.05	0.00	0.05	0.10	0.15	1.00
$\vec{\omega}_{\mathcal{K}_3}^{E,2}$	0.38	0.00	0.10	0.09	0.21	0.10	0.06	0.06	1.00
$\vec{\omega}_{\mathcal{K}_4}^{E,2}$	0.55	0.00	0.03	0.12	0.01	0.04	0.05	0.20	1.00
$\vec{\omega}_{\mathcal{K}_5}^{E,2}$	0.13	0.22	0.20	0.26	0.02	0.00	0.16	0.02	1.00
$\vec{\omega}_{\mathcal{K}_1}^{E,\infty}$	0.29	0.06	0.00	0.35	0.07	0.00	0.09	0.14	1.00
$\vec{\omega}_{\mathcal{K}_2}^{E,\infty}$	0.25	0.30	0.15	0.00	0.05	0.00	0.05	0.20	1.00
$\vec{\omega}_{\mathcal{K}_3}^{E,\infty}$	0.25	0.13	0.19	0.00	0.31	0.00	0.00	0.12	1.00
$\vec{\omega}_{\mathcal{K}_4}^{E,\infty}$	0.48	0.12	0.10	0.00	0.00	0.00	0.06	0.24	1.00
$\vec{\omega}_{\mathcal{K}_5}^{E,\infty}$	0.00	0.34	0.35	0.12	0.04	0.00	0.12	0.05	1.00
$\vec{\omega}_{\mathcal{K}_1}^{u}$	0	0.35	0.29	0.06	0	0	0.16	0.14	1.0
$\vec{\omega}_{\mathcal{K}_2}^{u}$	0.25	0.30	0.15	0	0	0.05	0.10	0.15	1.0
$\vec{\omega}_{\mathcal{K}_3}^{u}$	0.16	0.22	0.16	0.03	0.31	0	0.12	0	1.0
$\vec{\omega}_{\mathcal{K}_4}^{u}$	0.20	0.12	0.38	0	0.06	0.22	0	0.02	1.0
$\vec{\omega}_{\mathcal{K}_5}^{u}$	0.30	0.04	0	0.46	0	0	0.20	0	1.0

5.3 THE PROBLEMS WITH DISTANCE-BASED INTEGRATING PROBA-BILISTIC KNOWLEDGE BASES

The problem of probabilistic knowledge integration when each knowledge base determines a single probability function has been studied by Adamcik [29]. The solution for this problem is to combine both the geometrical notion of projections by means of a Bregman divergence and the framework of pooling operators. However, there are restrictions on the input PKBs, that is, they must be structurally similar and represented by the probability functions on a propositional language.

In our presentation, formally, we solve the problem of probabilistic knowledge integration in which the PKBs are represented by probabilistic constraint in many different structures. We effectively employ the DDFs in Table 5.2 and Table 5.3 for building and solving the problems with distance-based integrating PKBs.

Definition 5.5 represents a probability integrating vector of a PKB profile over a set of events.

Definition 5.5 ([51]) *(A probability integrating vector)*

Let $\mathcal{R} = \langle \mathcal{B}, \mathsf{E} \rangle$ be a PKB profile. A probability integrating vector of $\mathcal{R}$ over E is $\vec{\varpi}_{\mathcal{B}}^{\mathsf{E}} = (\mathcal{P}(\Theta_1), \ldots, \mathcal{P}(\Theta_{\hbar_E}))$ such that $\sum_{i=1}^{\hbar_E} \mathcal{P}(\Theta_i) = 1$.

Definition 5.6 represents a probability integrating function of a PKB profile over a set of events.

Definition 5.6 ([51]) *(A probability integrating function) Let $\mathcal{R} = \langle \mathcal{B}, \mathsf{E} \rangle$ be a PKB profile. A probability integrating function of $\mathcal{R}$ over E is a mapping $f : \mathbb{P}^{\hbar_E} \to \mathbb{R}^*$ such that $\forall \mathcal{K}_i \in \mathcal{B}: \mathcal{K}_i \not\models \bot$ and $\vec{\omega}_{\mathcal{K}_i}^{\mathsf{E}}$ is the satisfying probability vector of $\mathcal{K}_i$:*

$$f^{\vartheta}(\vec{y}) = \sum_{i=1}^{\hbar_{\mathcal{B}}} d_i^{\vartheta}(\vec{\omega}_{\mathcal{K}_i}^{\mathsf{E}}, \vec{y}) \tag{5.3}$$

where $d_i^{\vartheta}(\vec{\omega}_{\mathcal{K}_i}^{\mathsf{E}}, \vec{y})$ is a DDF from $\mathcal{K}_i$ to $\vec{y}$.

The following theorem shows that the probability integrating vector of a PKB profile over a set of events could be found by solving the non-linear optimization problem.

Theorem 5.8 ([51]) *Let $f^{\vartheta} : \mathbb{P}^{\hbar_E} \to \mathbb{R}^*$ be a probability integrating function $\mathcal{R}$ over E. The probability integrating vector $\vec{\varpi}_{\mathcal{B}}^{\mathsf{E}}$ of $\mathcal{R}$ over E is the optimal solution $\vec{y}^{\vartheta*}$ of* *the following non-linear optimization problem:*

$$\arg \min_{\vec{y} \in \mathbb{P}^{\hbar_E}} \sum_{i=1}^{\hbar_B} d_i^{\vartheta}(\vec{\omega}_{\mathcal{K}_i}^E, \vec{y}) \tag{5.4}$$

subject to $\vec{y} \in C_m$

where,

$C_m = \left\{ \vec{y} \in \mathbb{P}^{\hbar_E} | \sum_{j=1}^{\hbar_E} y_j = 1, y_j \geq 0 \, \forall j = \overline{1, \hbar_E} \right\}$ is a constraint set of the problem with the distance-based integrating PKBs.

Proof 5.8 *(a) Firstly, we prove that the problem (5.4) always exists a feasible solution. Constraint set C_m consists of $\hbar_E$ the linear functions. Since the DDFs introduced in Table 5.2 and Table 5.3 are both sums of linear, logarithmic, exponential, square, absolute value functions, the DDFs are all convex functions. As the objective function $f^{\vartheta}(\vec{y})$ is sums of DDFs, $f^{\vartheta}(\vec{y})$ is a convex function [68, 56]. Hence, the problem (5.4) is a convex optimization problem. Therefore, the problem (5.4) has a feasible solution $\vec{y}^{\vartheta*} \in C_m$ and attains optimal value $f^{\vartheta*} = \sum_{i=1}^{\hbar_B} d_i^{\vartheta}(\vec{\omega}_{\mathcal{K}_i}^E, \vec{y}^{\vartheta*})$.*

For instance, applying the DDF d^S in Table 5.2 simply returns $d^S(\vec{\omega}_{\mathcal{K}_i}^E, \vec{y}) = \sum_{j=1}^{\hbar_E}(p_{ij} - y_j)^2$ for each $\mathcal{K}_i$. Now assume that $\mathcal{F}$ is a solution of the problem (5.4) then there exists $\vec{y}^{\vartheta*}$ with $\mathcal{F} = \sum_{i=1}^{\hbar_B} d^S(\vec{\omega}_{\mathcal{K}_i}^E, \vec{y}^{\vartheta*}) = \sum_{i=1}^{\hbar_B} \sum_{j=1}^{\hbar_E}(p_{ij} - y_j)^2$ and $\vec{y}^{\vartheta*}$ is an optimal point.

Since the minimum of every convex function on a convex set is uniquely, $\vec{y}^{\vartheta*}$ is unique feasible and $f^{\vartheta*}$ is a uniquely optimal solution [68]. Therefore, $\vec{y}^{\vartheta*}$ is then optimal solution of the problem (5.4) [68].

(b) Secondly, we prove that $\vec{y}^{\vartheta*}$ is a probability integrating vector of $\mathcal{R}$ over E. As $\vec{y}^{\vartheta*} \in \mathbb{P}^{\hbar_E}$, $\vec{y}^{\vartheta*}$ satisfies the condition $\sum_{i=1}^{\hbar_E} \mathcal{P}(\Theta_i) = 1$ with $\vec{\omega}_B = \{\mathcal{P}(\Theta_1), \ldots, \mathcal{P}(\Theta_{\hbar_E})\}$. Therefore, using the definition of a probability integrating vector (Definition 5.5), $\vec{y}^{\vartheta*}$ is a probability integrating vector of $\mathcal{R}$ over E. □

The following theorem shows that the probability rules could be employed to find a corresponding consistent PKB from a probability integrating vector.

Theorem 5.9 ([51]) *Let $\mathcal{R} = \langle \mathcal{B}, E \rangle$ be a PKB profile such that $\forall \mathcal{K}_i \in \mathcal{B} : \mathcal{K}_i \nvDash \bot$. If there exists a probability integrating vector $\vec{\omega}_{\mathcal{B}}^E$ then there exists a PKB $\mathcal{K} = \varphi_{\mathcal{K}}^E(\vec{\omega}_{\mathcal{B}}^E)$ such that $\mathcal{K} \nvDash \bot$.*

Proof 5.9 *For the best case, PKBs have the same structure, that is, $\forall \mathcal{K}_i, \mathcal{K}_j$ $(i \neq j) \in \mathcal{B} : \bar{b}_{\mathcal{K}_i} = \bar{b}_{\mathcal{K}_j} = h$ and $\forall (F_k|G_k)[\rho_k] \in \mathcal{K}_i$, $(F_{k'}|G_{k'})[\rho_{k'}] \in \mathcal{K}_j$, $k = k' : (F_k|G_k) \simeq (F_{k'}|G_{k'})$. It follows that $\bar{b}_{\mathcal{K}} = h$.*

For the worst case, a PKB is completely different from all the remaining PKBs about structure, that is, $\forall \mathcal{K}_i, \mathcal{K}_j$ $(i \neq j) \in \mathcal{B} : \bar{b}_{\mathcal{K}_i} \neq \bar{b}_{\mathcal{K}_j}$ and $\forall (F_k|G_k)[\rho_k] \in \mathcal{K}_i$, $(F_{k'}|G_{k'})[\rho_{k'}] \in \mathcal{K}_j$, $k = k' : (F_k|G_k) \not\simeq (F_{k'}|G_{k'})$. It follows that $\bar{b}_{\mathcal{K}} = \sum_{i=1}^{h_{\mathcal{B}}} \bar{b}_{\mathcal{K}_i}$.

Let $t = \bar{b}_{\mathcal{K}}$.

For each constraint $(F)[\rho] \in \mathcal{K}$, based on rule (P0) in Theorem 2.1,

$$\rho = \sum_{\Theta \in \Lambda(E):\Theta \models F} \mathcal{P}(\Theta).$$

For each constraint $(F|G)[\rho] \in \mathcal{K}$, based on rule (P2) in Theorem 2.1,

$$\rho = \frac{\sum_{\Theta \in \Lambda(E):\Theta \models FG} \mathcal{P}(\Theta)}{\sum_{\Theta \in \Lambda(E):\Theta \models G} \mathcal{P}(\Theta)}.$$

Using the definition of the probabilistic inference function (Definition 4.1) and a probability integrating vector (Definition 5.5), we have

$$\mathcal{K} = \varphi_{\mathcal{K}}^{\mathsf{E}}(\vec{\varpi}_{\mathcal{B}}^{\mathsf{E}}) = \{(F_1|G_1)[\rho_1], \dots, (F_t|G_t)[\rho_t]\} \text{ and } \mathcal{K} \not\models \perp.$$ □

5.4 DISTANCE-BASED INTEGRATING OPERATORS

5.4.1 The Class of Probabilistic Integrating Operators Γ^ϑ

The following definition is employed to obtain a representation of the probabilistic integrating operators in terms of DDFs.

Definition 5.7 ([51, 88]) *(A probabilistic integrating operator Γ^ϑ) Let $\mathcal{R} = \langle \mathcal{B}, \mathsf{E} \rangle$ be a PKB. A probabilistic integrating operator $\Gamma : \mathcal{B} \to \mathbb{P}^{h_{\mathsf{E}}}$ of $\mathcal{B}$ over E in terms of DDF d^ϑ is defined as follows:*

$$\Gamma^\vartheta(\mathcal{B}) = \vec{\varpi}_{\mathcal{B}}^{\mathsf{E}} \tag{5.5}$$

We now state some theorems (5.10 and 5.11) relating the logical relationship between the desirable properties and the probabilistic integrating operators in terms of DDF.

Theorem 5.10 ([51, 88]) *The probabilistic integrating operators Γ^ϑ based on the DDFs, such as S-Div, M-Div, V-Div, K-Div, B-Div, Q-Div, H-Div, NA-Div, NAI-Div, D-Div, C-Div and R-Div satisfies all the desirable properties.*

Proof 5.10 *Consider the DDF S-Div given in Table 5.2.*

(*SCP*). *Firstly, assume that* $\vec{y}_{\mathcal{B}}^{S*} \in \Gamma^S(\mathcal{B})$. *From the definition of a probabilistic integrating operator (Definition 5.7),* $\vec{y}_{\mathcal{B}}^{S*}$ *is an optimal value of Problem (5.4) such that* $\vec{y}_{\mathcal{B}}^{S*} \in \mathbb{P}^{\hbar_E}$. *From the definition of a probability integrating function (Definition 5.6), there exists* $\vec{\omega}_{\mathcal{K}_i}^E \forall i = \overline{1, \hbar_{\mathcal{B}}}$. *Then,* $\exists \mathcal{P} \in \mho(\mathcal{K}_i) \forall i = \overline{1, \hbar_{\mathcal{B}}}$ *and the fact that* $\cap_{i=1}^{\hbar_{\mathcal{B}}} \mho(\mathcal{K}_i) \neq \emptyset$, $\exists \mathcal{P} \in \cap_{i=1}^{\hbar_{\mathcal{B}}} \mho(\mathcal{K}_i)$. *Therefore, By Definition 2.2,* $\vec{y}^S \in \cap_{i=1}^{\hbar_{\mathcal{B}}} \mho(\mathcal{K}_i)$.

Conversely, assume that $\vec{y}_{\mathcal{B}}^{S*} \in \cap_{i=1}^{\hbar_{\mathcal{B}}} \mho(\mathcal{K}_i)$. *Then, there exists* $\vec{y}^S \in \Gamma^S(\mathcal{K}_i) \forall i = \overline{1, \hbar_{\mathcal{B}}}$. *By Theorem 5.8,* $\vec{y}_{\mathcal{B}}^{S*} \in \Gamma^S(\mathcal{B})$.

(*CP*). *It is easy to draw from the property* **SCP**.

(*EIP*) *and* (*EP*). *Two properties are inferred from the definition of a probabilistic integrating operator (Definition 5.7).*

(*AP*). *Firstly, we need to prove that* $\Gamma^S(\mathcal{B}) \cap \Gamma^S(\mathcal{C}) \subseteq \Gamma^S(\mathcal{B} \cup \mathcal{C})$. *Since* $\Gamma^S(\mathcal{B}) \cap \Gamma^S(\mathcal{C}) \neq \emptyset$, *there exists* $\vec{y}^{S*} \in \Gamma^S(\mathcal{B}) \cap \Gamma^S(\mathcal{C})$. *Then, by Definition 5.7, Problem (5.4) has an optimal solution* $\vec{y}^{S*}$ *such that* $\vec{y}^{S*} \in \mathbb{P}^{\hbar_E}$, $\vec{y}^{S*} = \vec{y}_{\mathcal{B}}^{S*} = \vec{y}_{\mathcal{C}}^{S*}$. *Corresponding to* $\vec{y}^{S*}$, *there exists* $\vec{\alpha} = (\vec{\alpha}_1, \ldots, \vec{\alpha}_{\hbar_{\mathcal{B}}}) \in \Gamma^S(\mathcal{B})$ *and* $\vec{\beta} = (\vec{\beta}_1, \ldots, \vec{\beta}_{\hbar_{\mathcal{C}}}) \in \Gamma^S(\mathcal{C})$.

Using Theorem 5.8:

- *Problem (5.4) has an optimal value* $f^{\vartheta*}(\vec{y}_{\mathcal{B}}^{S*})$, *where*

$$f^{\vartheta*}(\vec{y}_{\mathcal{B}}^{S*}) = \sum_{i=1}^{\hbar_{\mathcal{B}}} d^S(\vec{\alpha}_i, \vec{y}_{\mathcal{B}}^{S*}) = \sum_{i=1}^{\hbar_{\mathcal{B}}} \sum_{j \in I(\vec{y}_{\mathcal{B}}^{S*})} (\alpha_{ij} - y_j)^2 \text{ with } \vec{\alpha}_i = \vec{\omega}_{\mathcal{K}_i}^E \text{ and } \vec{y}_{\mathcal{B}}^{S*} \in \mathbb{P}^{\hbar_E}$$

- *Problem (5.4) has an optimal value* $f^{\vartheta*}(\vec{y}_{\mathcal{C}}^{S*})$, *where*

$$f^{\vartheta*}(\vec{y}_{\mathcal{C}}^{S*}) = \sum_{i=1}^{\hbar_{\mathcal{C}}} d^{S*}(\vec{\beta}_i, \vec{y}_{\mathcal{C}}^{S*}) = \sum_{i=1}^{\hbar_E} \sum_{j \in I(\vec{y}_{\mathcal{C}}^{S*})} (\beta_{ij} - y_j)^2 \text{ with } \vec{\beta}_i = \vec{\omega}_{\mathcal{K}'_i}^E \text{ and } \vec{y}_{\mathcal{C}}^{S*} \in \mathbb{P}^{\hbar_E}.$$

It follows that $f^{\vartheta*}(\vec{y}_{\mathcal{B}}^{S*}) + f^{\vartheta*}(\vec{y}_{\mathcal{C}}^{S*}) \leq f^{\vartheta*}(\vec{y}_{\mathcal{B} \cup \mathcal{C}}^{S*})$ *and it can be seen that* $\vec{y}^{S*}, \vec{\alpha}, \vec{\beta}$ *is a global minimum of* $\sum_{i=1}^{\hbar_{\mathcal{B}}} \sum_{j \in I(\vec{y}_{\mathcal{B}}^{t*})} (\alpha_{ij} - y_j)^2 + \sum_{i=1}^{\hbar_E} \sum_{j \in I(\vec{y}_{\mathcal{C}}^{t*})} (\beta_{ij} - y_j)^2$ *with* $\vec{\alpha}_i = \vec{\omega}_{\mathcal{K}_i}^E \forall i = \overline{1, \hbar_{\mathcal{B}}}$ *and* $\vec{\beta}_i = \vec{\omega}_{\mathcal{K}'_i}^E \forall i = \overline{1, \hbar_{\mathcal{C}}}$.

Thus, $\vec{y}^{S*} \in \Gamma^S(\mathcal{B} \cup \mathcal{C})$ *and* $f^{\vartheta*}(\vec{y}_{\mathcal{B}}^{S*}) + f^{\vartheta*}(\vec{y}_{\mathcal{C}}^{S*}) = f^{\vartheta*}(\vec{y}_{\mathcal{B} \cup \mathcal{C}}^{S*})$.

Therefore, $\Gamma^S(\mathcal{B}) \cap \Gamma^S(\mathcal{C}) \subseteq \Gamma^S(\mathcal{B} \cup \mathcal{C})$.

Secondly, we prove that $\Gamma^S(\mathcal{B} \cup \mathcal{C}) \subseteq \Gamma^S(\mathcal{B}) \cap \Gamma^S(\mathcal{C})$.

Assume that $\vec{y} \in \Gamma^S(\mathcal{B} \cup \mathcal{C})$. *Then,* $\vec{y} = \vec{y}_{\mathcal{B} \cup \mathcal{C}}^{S*}$ *and there exists* $(\vec{\alpha}_1, \ldots, \vec{\alpha}_{\hbar_{\mathcal{B}}}, \vec{\beta}_1, \ldots, \vec{\beta}_{\hbar_{\mathcal{C}}})$ *such that* $f^{\vartheta*}(\vec{y}_{\mathcal{B} \cup \mathcal{C}}^{S*}) = \sum_{i=1}^{\hbar_{\mathcal{B}}} \sum_{j \in I(y)} (\alpha_{ij} - y_j)^2 + \sum_{i=1}^{\hbar_E} \sum_{j \in I(y)} (\beta_{ij} - y_j)^2$ *is an optimal value of Problem (5.4).*

Assume that there is not a simultaneous existence of $\vec{y}_{\mathcal{B}}^{S*}$ *such that* $f^{\vartheta*}(\vec{y}_{\mathcal{B}}^{S*}) = \sum_{i=1}^{\hbar_{\mathcal{B}}} \sum_{j \in I(y)} (\alpha_{ij} - y_j)^2$ *and* $\vec{y}_{\mathcal{C}}^{S*}$ *such that* $f^{\vartheta}(\vec{y}_{\mathcal{C}}^{S*}) = \sum_{i=1}^{\hbar_E} \sum_{j \in I(y)} (\beta_{ij} - y_j)^2$. *This is in contrast with finding the minimum of the function* $f^{\vartheta}(\vec{y}_{\mathcal{B}}^S)$ *or* $f^{\vartheta}(\vec{y}_{\mathcal{C}}^S)$. *It follows that* $\vec{y} \in \Gamma^S(\mathcal{B}) \cap \Gamma^S(\mathcal{C})$. *Therefore,* $\Gamma^S(\mathcal{B} \cup \mathcal{C}) \subseteq \Gamma^S(\mathcal{B}) \cap \Gamma^S(\mathcal{C})$.

(SDP). Assume that $\Gamma(\mathcal{B} \cup \mathcal{C}) \cap \Gamma(\mathcal{B}) \neq \emptyset$, so that $\vec{y}^{S*} \in \Gamma^S(\mathcal{B} \cup \mathcal{C})$ and $\vec{y}^{S*} \in \Gamma^S(\mathcal{B})$. Thus, there exist

- *$\vec{y}^{S*} = \vec{y}_{\mathcal{B} \cup \mathcal{C}}^{S*}$, $(\vec{\alpha}_1, \ldots, \vec{\alpha}_{\hbar_B}$ and $\vec{\beta}_1, \ldots, \vec{\beta}_{\hbar_C})$ such that $f^{\vartheta*}(\vec{y}_{\mathcal{B} \cup \mathcal{C}}^{S*}) = \sum_{i=1}^{\hbar_B} \sum_{j \in I(y)} (\alpha_{ij} - y_j)^2 + \sum_{i=1}^{\hbar_E} \sum_{j \in I(y)} (\beta_{ij} - y_j)^2$ is an optimal value of Problem (5.4)*

- *$\vec{y}^{S*} = \vec{y}_{\mathcal{B}}^{S*}$, $(\vec{\alpha}_1, \ldots, \vec{\alpha}_{\hbar_B})$ such that $f^{\vartheta*}(\vec{y}_{\mathcal{B}}^{S*}) = \sum_{i=1}^{\hbar_B} \sum_{j \in I(y)} (\alpha_{ij} - y_j)^2$ is an optimal value of Problem (5.4).*

As $\Gamma(\mathcal{B}) \cap \Gamma(\mathcal{C}) \neq \emptyset$, $\vec{y}^{S*} \in \Gamma^S(\mathcal{B})$ and $\vec{y}^{S*} \in \Gamma^S(\mathcal{C})$ do not exist, which contradicts the assumption. It follows that $\Gamma(\mathcal{B} \cup \mathcal{C}) \cap \Gamma(\mathcal{B}) = \emptyset$.

*(PIP). Using Property **SCP**, as $\cap_{i=1}^{\hbar_C} \mho(\mathcal{K}'_i) \neq \emptyset$, $\Gamma^S(\mathcal{C}) = \cap_{i=1}^{\hbar_C} \mho(\mathcal{K}'_i)$. Hence, $\Gamma^S(\mathcal{B}) \subseteq \Gamma^S(\mathcal{C})$. It follows that $\Gamma^S(\mathcal{B}) \cap \Gamma^S(\mathcal{C}) = \Gamma^S(\mathcal{B})$. Moreover, using Property **AP**, $\Gamma^S(\mathcal{B}) \cap \Gamma^S(\mathcal{C}) = \Gamma^S(\mathcal{B} \cup \mathcal{C})$. Therefore, $\Gamma^S(\mathcal{B}) = \Gamma^S(\mathcal{B} \cup \mathcal{C})$.*

*As the DDFs **M**-Div, **V**-Div, **K**-Div, **B**-Div, **Q**-Div, and **H**-Div in Table 5.2 are all derived from basic mathematical formulas. It follows that the proof of the properties of $\Gamma^M(\mathcal{B})$, $\Gamma^V(\mathcal{B})$, $\Gamma^K(\mathcal{B})$, and $\Gamma^B(\mathcal{B})$ will be obtained by applying the proof of the properties of Γ^S, where the DDF $d^S(\vec{x}, \vec{y})$ is replaced by the corresponding DDFs.*

*As the DDFs **NA**-Div, **NAI**-Div, **D**-Div, **C**-Div, **R**-Div in Table 5.3 are constructed with respect to the coefficient. It follows that the proof of the properties of $\Gamma^{NA}(\mathcal{B})$, $\Gamma^{NAI}(\mathcal{B})$, $\Gamma^D(\mathcal{B})$, $\Gamma^C(\mathcal{B})$, $\Gamma^R(\mathcal{B})$ is similar to the proof of the properties of Γ^S, where the DDF $d^S(\vec{x}, \vec{y})$ is replaced by the corresponding DDFs.* □

Theorem 5.11 ([51, 88]) *The probabilistic integrating operators Γ^ϑ based on the DDFs, such as **A**-Div, **AI**-Div, **C**-Div, **L**-Div, **J**-Div, **N**-Div, **I**-Div, **T**-Div and **E**-Div satisfies the desirable properties: **CP**, **SCP**, **EIP**, **EP**, **PIP** and **AP**.*

Proof 5.11 *Consider the DDF **L**-Div given in Table 5.2.*

(AP). Firstly, we prove that $\Gamma^S(\mathcal{B}) \cap \Gamma^S(\mathcal{C}) \subseteq \Gamma^S(\mathcal{B} \cup \mathcal{C})$. As $\Gamma^S(\mathcal{B}) \cap \Gamma^S(\mathcal{C}) \neq \emptyset$, there exists $\vec{y}^{L*} \in \Gamma^S(\mathcal{B}) \cap \Gamma^S(\mathcal{C})$. Then, by Definition 5.7, $\vec{y}^{L*}$ is an optimal value of Problem (5.4) such that $\vec{y}^{L*} \in \mathbb{P}^{\hbar_E}$, $\vec{y}^{L*} = \vec{y}_{\mathcal{B}}^{L*} = \vec{y}_{\mathcal{C}}^{L*}$. Corresponding to $\vec{y}^{L*}$ then there exist $\vec{\alpha} = (\vec{\alpha}_1, \ldots, \vec{\alpha}_{\hbar_B}) \in \Gamma^S(\mathcal{B})$ and $\vec{\beta} = (\vec{\beta}_1, \ldots, \vec{\beta}_{\hbar_C}) \in \Gamma^S(\mathcal{C})$.

By Theorem 5.8,

- *$f^\vartheta(\vec{y}_{\mathcal{B}}^{L*})$ is an optimal value of Problem (5.4), where*
$f^\vartheta(\vec{y}_{\mathcal{B}}^{L*}) = \sum_{i=1}^{\hbar_B} d^L(\vec{\alpha}_i, \vec{y}_{\mathcal{B}}^{L*}) = \sum_{i=1}^{\hbar_B} \sum_{j \in I(y)} \alpha_{ij} log \frac{\alpha_{ij}}{y_j}$ with $\vec{\alpha}_i = \vec{\omega}_{\mathcal{K}_i}^E$ and $\vec{y}_{\mathcal{B}}^{L*} \in \mathbb{P}^{\hbar_E}$
- *$f^\vartheta(\vec{y}_{\mathcal{C}}^{L*})$ is an optimal value of Problem (5.4), where*
$f^\vartheta(\vec{y}_{\mathcal{C}}^{L*}) = \sum_{i=1}^{\hbar_C} d^L(\vec{\beta}_i, \vec{y}_{\mathcal{C}}^{L*}) = \sum_{i=1}^{\hbar_E} \sum_{j \in I(y)} \beta_{ij} log \frac{\beta_{ij}}{y_j}$ with $\vec{\beta}_i = \vec{\omega}_{\mathcal{K}'_i}^E$ and $\vec{y}_{\mathcal{C}}^{L*} \in \mathbb{P}^{\hbar_E}$.

It follows that $f^{\vartheta}(\vec{y}_{\mathcal{B}}^{L*}) + f^{\vartheta}(\vec{y}_{\mathcal{C}}^{L*}) \leq f^{\vartheta *}(\vec{y}_{\mathcal{B}\cup\mathcal{C}}^{L})$, and it can be seen that $\vec{y}^{L*}, \vec{\alpha}, \vec{\beta}$ is a global minimum of $\sum\limits_{i=1}^{\hbar_{\mathcal{B}}} \sum\limits_{j\in I(y)} \alpha_{ij} log\frac{\alpha_{ij}}{y_j} + \sum\limits_{i=1}^{\hbar_E} \sum\limits_{j\in I(y)} \beta_{ij} log\frac{\beta_{ij}}{y_j}$ with $\vec{\alpha}_i = \vec{\omega}_{\mathcal{K}_i}^{E} \forall i = \overline{1, \hbar_{\mathcal{B}}}$ and $\vec{\beta}_i = \vec{\omega}_{\mathcal{K}'_i}^{E} \forall i = \overline{1, \hbar_{\mathcal{C}}}$.

It follows that $\vec{y}^{L*} \in \Gamma^L(\mathcal{B}\cup\mathcal{C})$. If $f^{\vartheta}(\vec{y}_{\mathcal{B}}^{L*})$ and $f^{\vartheta}(\vec{y}_{\mathcal{C}}^{L*})$ are equal to $+\infty$ then $f^{\vartheta *}(\vec{y}_{\mathcal{B}}^{L*}) + f^{\vartheta *}(\vec{y}_{\mathcal{C}}^{L}) = f^{\vartheta *}(\vec{y}_{\mathcal{B}\cup\mathcal{C}}^{L})$.

Therefore, $\Gamma^L(\mathcal{B}) \cap \Gamma^L(\mathcal{C}) \subseteq \Gamma^L(\mathcal{B}\cup\mathcal{C})$.

Secondly, we prove that $\Gamma^L(\mathcal{B}\cup\mathcal{C}) \subseteq \Gamma^L(\mathcal{B}) \cap \Gamma^L(\mathcal{C})$.

Assume that $\vec{y} \in \Gamma^L(\mathcal{B}\cup\mathcal{C})$. Then, $\vec{y} = \vec{y}_{\mathcal{B}\cup\mathcal{C}}^{L*}$ and there exists $(\vec{\alpha}_1, \ldots, \vec{\alpha}_{\hbar_{\mathcal{B}}}, \vec{\beta}_1, \ldots, \vec{\beta}_{\hbar_{\mathcal{C}}})$ such that $f^{\vartheta *}(\vec{y}_{\mathcal{B}\cup\mathcal{C}}^{L*}) = \sum\limits_{i=1}^{\hbar_{\mathcal{B}}} \sum\limits_{j\in I(\vec{y}_{\mathcal{B}}^{L*})} \alpha_{ij} log\frac{\alpha_{ij}}{y_j} +$ $\sum\limits_{i=1}^{\hbar_E} \sum\limits_{j\in I(\vec{y}_{\mathcal{C}}^{L*})} \beta_{ij} log\frac{\beta_{ij}}{y_j}$ is an optimal value of Problem (5.4).

Assume that there is not a simultaneous existence of $f^{\vartheta *}(\vec{y}_{\mathcal{B}}^{L*}) = \sum_{i=1}^{\hbar_{\mathcal{B}}} \sum\limits_{j\in I(\vec{y}_{\mathcal{B}}^{L*})} \alpha_{ij} log\frac{\alpha_{ij}}{y_j} \neq +\infty$ and $f^{\vartheta}(\vec{y}_{\mathcal{C}}^{L}) = \sum_{i=1}^{\hbar_E} \sum\limits_{j\in I(\vec{y}_{\mathcal{C}}^{L*})} \beta_{ij} log\frac{\beta_{ij}}{y_j} \neq +\infty$ which contradict $f^{\vartheta *}(\vec{y}_{\mathcal{B}}^{L*})$ or $f^{\vartheta *}(\vec{y}_{\mathcal{C}}^{L})$. It follows that $\vec{y} \in \Gamma^L(\mathcal{B}) \cap \Gamma^L(\mathcal{C})$. Therefore, $\Gamma^L(\mathcal{B}\cup\mathcal{C}) \subseteq \Gamma^L(\mathcal{B}) \cap \Gamma^L(\mathcal{C})$.

The proof of the remaining properties of Γ^L is similar to the proof of the properties of Γ^S where the DDF $d^S(\vec{x}, \vec{y})$ is replaced by $d^L(\vec{x}, \vec{y})$. $\square$

5.4.2 The Class of Probabilistic Integrating Operators Γ^{HU}

The following theorem shows that a probability integrating vector could be found by using a Formula (5.6).

Theorem 5.12 ([51, 88]) *Let $\mathcal{R} = \langle \mathcal{B}, E \rangle$ a PKB profile, $\mathcal{B} = \{\mathcal{K}_1, \ldots, \mathcal{K}_{\hbar_{\mathcal{B}}}\}$ and $\forall \mathcal{K}_i \in \mathcal{B}$ and $\mathcal{K}_i \not\models \perp$. Let $\hat{A}_{\mathcal{B}} = \left((\vec{\omega}_{\mathcal{K}_1}^{E})^T, \ldots, (\vec{\omega}_{\mathcal{K}_{\hbar_{\mathcal{B}}}}^{E})^T \right)$ be a profile matrix. There exist positive real numbers $\lambda_i \in [0, 1]$ with $\sum\limits_{i=1}^{\hbar_{\mathcal{B}}} \lambda_i = 1$ such that*

$$\vec{y}_{\mathcal{B}}^{HU} = \hat{A}_{\mathcal{B}} \vec{\lambda} \tag{5.6}$$

where $\vec{\lambda} = (\lambda_1, \ldots, \lambda_{\hbar_{\mathcal{B}}})^T$

Proof 5.12 *Let $\mathcal{B} = \{\mathcal{K}_1, \ldots, \mathcal{K}_{\hbar_{\mathcal{B}}}\}$ with $\vec{\omega}_{\mathcal{K}_i}^{E} = \{\rho_{i1}, \ldots, \rho_{i\hbar_E}\} \forall i = \overline{1, \hbar_{\mathcal{B}}}$. Thus,*

$$\vec{y}_{\mathcal{B}}^{HU} = \begin{pmatrix} \rho_{11} & \rho_{21} & \cdots & \rho_{\hbar_{\mathcal{B}}1} \\ \cdots & \cdots & \cdots & \cdots \\ \rho_{1\hbar_E} & \rho_{2\hbar_E} & \cdots & \rho_{\hbar_{\mathcal{B}}\hbar_E} \end{pmatrix} \begin{pmatrix} \lambda_1 \\ \cdots \\ \lambda_{\hbar_{\mathcal{B}}} \end{pmatrix} = \begin{pmatrix} \rho_{11}\lambda_1 + \cdots + \rho_{\hbar_{\mathcal{B}}1}\lambda_{\hbar_{\mathcal{B}}} \\ \cdots \\ \rho_{1\hbar_E}\lambda_1 + \cdots + \rho_{\hbar_{\mathcal{B}}\hbar_E}\lambda_{\hbar_{\mathcal{B}}} \end{pmatrix}$$

Let $\mathcal{Y}_j = \rho_{1j}\lambda_1 + \cdots + \rho_{\hbar_\mathcal{B}j}\lambda_{\hbar_\mathcal{B}} \,\forall j = \overline{1,\hbar_E}$. As $\vec{y}_\mathcal{B}^{HU}$ is the set of all probability func-tions satisfying the resulting PKB, $\sum_{i=1}^{\hbar_E} \mathcal{Y}_j = 1$ and $0 \leq \mathcal{Y}_j \leq 1$. Hence, $\sum_{i=1}^{\hbar_\mathcal{B}} \lambda_i = 1$ and $0 \leq \lambda_i \leq 1 \,\forall i = \overline{1,\hbar_\mathcal{B}}$. Therefore, we obtain Formula (5.6). $\qquad\square$

The following definition is employed to obtain a representation of the probabilistic integrating operator built from a formula.

Definition 5.8 ([51, 88]) *(A probabilistic integrating operator Γ^{HU}). A probabilistic integrating operator $\Gamma^{HU} : \mathcal{B} \to \mathbb{P}^{\hbar_E}$ is defined by $\Gamma^{HU}(\mathcal{B}) = \vec{y}_\mathcal{B}^{HU}$.*

The following theorem presents the logical relationship between the desirable properties and the formula-based probabilistic integrating operator.

Theorem 5.13 ([88]) *A probabilistic integrating operator Γ^{HU} satisfies **CP**, **SCP**, **EIP**, **EP**, and **PIP**.*

Proof 5.13 *(**SCP**). Firstly, assume that $\vec{y} \in \Gamma^{HU}(\mathcal{B})$. By Definition 5.8, $\vec{y} \in \hat{A}_\mathcal{B}\vec{\lambda}$. As $\cap_{i=1}^{\hbar_C}\mho(\mathcal{K}_i) \neq \emptyset$, $\vec{y} \in \cap_{i=1}^{\hbar_C}\mho(\mathcal{K}_i)$.*

By contrast, assume that $\vec{y} \in \cap_{i=1}^{\hbar_C}\mho(\mathcal{K}_i)$. It follows that there exists $\vec{y} \in \Gamma^{HU}(\mathcal{K}_i)\forall i = \overline{1,\hbar_\mathcal{B}}$. Therefore, $\vec{y} \in \Gamma^{HU}(\mathcal{B})$ follows from Theorem 5.12.

*(**CP**). It is easy to see that it follows from **SCP**.*

*(**EIP**, and **EP**. Two properties are inferred from Definition 5.8.*

*(**PIP**). It is easy to attain from the properties **SCP** and **AP**.* $\qquad\square$

Table 5.6 summarizes the logical relationship between desirable properties and the probabilistic integrating operators.

Example 5.3 *This example describes the probabilistic knowledge integration with an input that is a PKB profile in Example 2.3.*

*   **Step 1**: Computing the inconsistency measures $\mathcal{I}(\mathcal{K}_1)$, $\mathcal{I}(\mathcal{K}_2)$, $\mathcal{I}(\mathcal{K}_3)$, $\mathcal{I}(\mathcal{K}_4)$, $\mathcal{I}(\mathcal{K}_5)$. The results are shown in Table 3.2.*

*   **Step 2**: Restoring the consistency of PKBs $\mathcal{K}_1$ and $\mathcal{K}_5$. The results are shown in Table 4.5*

*   **Step 3**: Find the satisfying probability vectors $\vec{\omega}_{\mathcal{K}_1}^E$ of $\mathcal{K}_1$, $\vec{\omega}_{\mathcal{K}_2}^E$ of $\mathcal{K}_2$, $\vec{\omega}_{\mathcal{K}_3}^E$ of $\mathcal{K}_3$, $\vec{\omega}_{\mathcal{K}_4}^E$ of $\mathcal{K}_4$ and $\vec{\omega}_{\mathcal{K}_5}^E$ of $\mathcal{K}_5$ over E. The results are shown in Table 5.5.*

*   **Step 4**: Integrating PKBs of $\mathcal{B}$.*

*   1) Consider the DDF **S-Div**. By Definition 5.6 and from Table 5.2, for 1-norm, we have the following DDFs:*

Table 5.6 The logical relationship between desirable properties and the probabilistic integrating operators [51]

Operators	CP	SCP	EIP	EP	PIP	AP	DP	SDP	Theorems
Γ^S	√	√	√	√	√	√	√	√	5.10
Γ^M	√	√	√	√	√	√	√	√	5.10
Γ^V	√	√	√	√	√	√	√	√	5.10
Γ^K	√	√	√	√	√	√	√	√	5.10
Γ^B	√	√	√	√	√	√	√	√	5.10
Γ^Q	√	√	√	√	√	√	√	√	5.10
Γ^H	√	√	√	√	√	√	√	√	5.10
Γ^A	√	√	√	√	√	√	-	-	5.11
Γ^{NA}	√	√	√	√	√	√	√	√	5.10
Γ^{AI}	√	√	√	√	√	√	-	-	5.11
Γ^{NAI}	√	√	√	√	√	√	√	√	5.10
Γ^C	√	√	√	√	√	√	√	√	5.10
Γ^D	√	√	√	√	√	√	√	√	5.11
Γ^{CS}	√	√	√	√	√	√	-	-	5.11
Γ^R	√	√	√	√	√	√	√	√	5.11
Γ^L	√	√	√	√	√	√	-	-	5.11
Γ^J	√	√	√	√	√	√	-	-	5.11
Γ^N	√	√	√	√	√	√	-	-	5.11
Γ^I	√	√	√	√	√	√	-	-	5.11
Γ^T	√	√	√	√	√	√	-	-	5.10
Γ^E	√	√	√	√	√	√	-	-	5.11
Γ^{HUL}	√	√	√	√	√	-	-	√	5.13

- $d_1^S(\vec{\omega}_{\mathcal{K}_1}^E, \vec{y}) = (-y_1)^2 + (0.35 - y_2)^2 + (0.29 - y_3)^2 + (0.06 - y_4)^2 + (0.10 - y_5)^2 + (-y_6)^2 + (0.06 - y_7)^2 + (0.14 - y_8)^2$

- $d_2^S(\vec{\omega}_{\mathcal{K}_2}^E, \vec{y}) = (0.25 - y_1)^2 + (0.30 - y_2)^2 + (0.15 - y_3)^2 + (-y_4)^2 + (0.05 - y_5)^2 + (-y_6)^2 + (0.05 - y_7)^2 + (0.20 - y_8)^2$

- $d_3^S(\vec{\omega}_{\mathcal{K}_3}^E, \vec{y}) = (0.25 - y_1)^2 + (0.13 - y_2)^2 + (0.19 - y_3)^2 + (-y_4)^2 + (0.31 - y_5)^2 + (-y_6)^2 + (-y_7)^2 + (0.12 - y_8)^2$

- $d_4^S(\vec{\omega}_{\mathcal{K}_4}^E, \vec{y}) = (0.54 - y_1)^2 + (-y_2)^2 + (0.04 - y_3)^2 + (0.12 - y_4)^2 + (0.06 - y_5)^2 + (-y_6)^2 + (-y_7)^2 + (0.24 - y_8)^2$

- $d_5^S(\vec{\omega}_{\mathcal{K}_5}^E, \vec{y}) = (-y_1)^2 + (0.34 - y_2)^2 + (0.35 - y_3)^2 + (0.11 - y_4)^2 + (0.05 - y_5)^2 + (-y_6)^2 + (0.10 - y_7)^2 + (0.05 - y_8)^2$

Using the definition of a probability integrating function (Definition 5.6), we have the probability integrating function

$$f^S(\vec{y}) = d_1^S(\vec{\omega}_{\mathcal{K}_1}^E, \vec{y}) + d_2^S(\vec{\omega}_{\mathcal{K}_2}^E, \vec{y}) + d_3^S(\vec{\omega}_{\mathcal{K}_3}^E, \vec{y}) + d_4^S(\vec{\omega}_{\mathcal{K}_4}^E, \vec{y}) + d_5^S(\vec{\omega}_{\mathcal{K}_5}^E, \vec{y}).$$

By Theorem 5.8, the probability integrating vector $\vec{y}^{S*}$ of $\mathcal{R}$ is the optimal solution of the following non-linear optimization problem:

$$\arg\min_{\vec{y}\in\mathbb{P}^8} f^S(\vec{y}) \tag{5.7}$$

subject to

$$y_1 + y_2 + y_3 + y_4 + y_5 + y_6 + y_7 + y_8 = 1 \tag{5.8}$$

$$y_1 \geq 0, y_2 \geq 0, y_3 \geq 0, y_4 \geq 0, y_5 \geq 0, y_6 \geq 0, y_7 \geq 0, y_8 \geq 0, \tag{5.9}$$

Problem (5.7) has the optimal solution

$\vec{y}_{\mathcal{B}}^{S*} = (0.21, 0.20, 0.22, 0.06, 0.12, 0.03, 0.01, 0.14).$

By Definition 5.7, $\Gamma^S(\mathcal{B}) = \vec{y}_{\mathcal{B}}^{S*} = (0.21, 0.20, 0.22, 0.06, 0.12, 0.03, 0.01, 0.14).$

The probability integrating vectors of $\mathcal{R}$ with respect to 1-norm corresponding to DDFs are shown in Table 5.7.

2) Consider the DDF $\mathbf{B}$-Div. By Definition 5.6 and from Table 5.2, for 2-norm, we have the followwing DDFs:

- $d_1^B(\vec{\omega}_{\mathcal{K}_1}^E, \vec{y}) = 2 - 2(\sqrt{0.14y_1} + \sqrt{0.21y_2} + \sqrt{0.14y_3} + \sqrt{0.21y_4} + \sqrt{0.05y_5} + \sqrt{0.04y_6} + \sqrt{0.11y_7} + \sqrt{0.10y_8})$

- $d_2^B(\vec{\omega}_{\mathcal{K}_2}^E, \vec{y}) = 2 - 2(\sqrt{0.30y_1} + \sqrt{0.25y_2} + \sqrt{0.10y_3} + \sqrt{0.05y_4} + \sqrt{0.05y_6} + \sqrt{0.10y_7} + \sqrt{0.15y_8})$

- $d_3^B(\vec{\omega}_{\mathcal{K}_3}^E, \vec{y}) = 2 - 2(\sqrt{0.38y_1} + \sqrt{0.11y_3} + \sqrt{0.07y_4} + \sqrt{0.17y_5} + \sqrt{0.13y_6} + \sqrt{0.08y_7} + \sqrt{0.04y_8})$

Table 5.7 The probability integrating vectors of $\mathcal{R}$ with respect to 1-norm [51]

Vectors	λ	HTD	$HT\bar{D}$	$H\bar{T}D$	$H\bar{T}\bar{D}$	$\bar{H}TD$	$\bar{H}T\bar{D}$	$\bar{H}\bar{T}D$	$\bar{H}\bar{T}\bar{D}$
Γ^A	0.50	0.16	0.24	0.22	0.05	0.14	0.02	0.00	0.17
Γ^C	2.00	0.22	0.19	0.22	0.06	0.12	0.04	0.02	0.13
Γ^{CS}	0.70	0.00	0.33	0.00	0.13	0.19	0.09	0.03	0.23
Γ^{NA}	0.80	0.19	0.21	0.23	0.05	0.13	0.03	0.01	0.15
Γ^{NAI}	0.90	0.03	0.06	0.06	0.06	0.10	0.62	0.05	0.02
Γ^D	0.90	0.20	0.23	0.24	0.05	0.12	0.03	0.00	0.13
Γ^R	2.00	0.21	0.20	0.22	0.06	0.12	0.03	0.01	0.14
Γ^{AI}	1.20	0.22	0.22	0.24	0.05	0.12	0.02	0.00	0.13
Γ^N		0.21	0.20	0.23	0.06	0.12	0.03	0.01	0.14
Γ^E		0.24	0.18	0.21	0.07	0.11	0.05	0.02	0.11
Γ^I		0.18	0.22	0.23	0.05	0.13	0.02	0.00	0.16
Γ^J		0.29	0.17	0.19	0.06	0.09	0.07	0.02	0.10
Γ^L		0.21	0.20	0.22	0.06	0.12	0.03	0.01	0.14
Γ^T		0.26	0.18	0.16	0.04	0.08	0.0	0.09	0.17
Γ^B		0.16	0.24	0.22	0.05	0.14	0.02	0.00	0.17
Γ^H		0.23	0.22	0.20	0.04	0.13	0.02	0.00	0.16
Γ^Q		0.23	0.22	0.20	0.04	0.13	0.02	0.00	0.16
Γ^K		0.23	0.19	0.21	0.06	0.13	0.04	0.02	0.12
Γ^S		0.21	0.20	0.22	0.06	0.12	0.03	0.01	0.14
Γ^V		0.24	0.18	0.30	0.06	0.09	0.00	0.00	0.13
Γ^M		0.20	0.21	0.24	0.06	0.12	0.03	0.00	0.15
Γ^{HU}		0.29	0.17	0.16	0.05	0.14	0	0.03	0.17

$- d_4^B(\vec{\omega}_{\mathcal{K}_4}^E, \vec{y}) = 2 - 2(\sqrt{0.55y_1} + \sqrt{0.03y_3} + \sqrt{0.12y_4} + \sqrt{0.01y_5} + \sqrt{0.04y_6} + \sqrt{0.05y_7} + \sqrt{0.19y_8})$

$d_5^B(\vec{\omega}_{\mathcal{K}_5}^E, \vec{y}) = 2 - 2(\sqrt{0.13y_1} + \sqrt{0.22y_2} + \sqrt{0.20y_3} + \sqrt{0.26y_4} + \sqrt{0.02y_5} + \sqrt{0.16y_7} + \sqrt{0.02y_8}).$

Using the definition of a probability integrating function (Definition 5.6), we have the probability integrating function $f^B(\vec{y}) = d_1^B(\vec{\omega}_{\mathcal{K}_1}^E, \vec{y}) + d_2^B(\vec{\omega}_{\mathcal{K}_2}^E, \vec{y}) + d_3^B(\vec{\omega}_{\mathcal{K}_3}^E, \vec{y}) + d_4^B(\vec{\omega}_{\mathcal{K}_4}^E, \vec{y}) + d_5^B(\vec{\omega}_{\mathcal{K}_5}^E, \vec{y}).$ *By Theorem 5.8, the probability integrating vector* $\vec{y}^{B*}$ *of* $\mathcal{R}$ *is the optimal solution of the following non-linear optimization problem:*

$$\arg \min_{\vec{y} \in \mathbb{P}^8} f^B(\vec{y}) \qquad (5.10)$$

subject to (5.8-5.9)

Problem (5.10) has the optimal solotion

$\vec{y}_{\mathcal{B}}^{B*} = (0.03, 0.06, 0.06, 0.06, 0.10, 0.62, 0.05, 0.02).$

By Definition 5.7, $\Gamma^B(\mathcal{B}) = \vec{y}_{\mathcal{B}}^{B*} = (0.03, 0.06, 0.06, 0.06, 0.10, 0.62, 0.05, 0.02).$ *The probability integrating vectors of* $\mathcal{R}$ *with respect to 2-norm corresponding to DDFs are shown in Table 5.8.*

3) Consider the DDF **R**-*Div. By Definition 5.6 and from Table 5.3, for* ∞-*norm and* $\lambda = 2$, *we have the followwing DDFs:*

$- d_1^R(\vec{\omega}_{\mathcal{K}_1}^E, \vec{y}) = 0.29^2 - y_1^2 - 2(0.29 - y_1)y_1 + 0^2 - y_2^2 - 2(0 - y_2)y_2 + 0.06^2 - y_3^2 - 2(0.06 - y_3)y_3 + 0.35^2 - y_4^2 - 2(0.35 - y_4)y_4 + 0.07^2 - y_5^2 - 2(0.07 - y_5)y_5 + 0.09^2 - y_6^2 - 2(0.09 - y_6)y_6 + 0^2 - y_7^2 - 2(0 - y_7)y_7 + 0.14^2 - y_8^2 - 2(0.14 - y_8)y_8$

$- d_2^R(\vec{\omega}_{\mathcal{K}_2}^E, \vec{y}) = +0.25^2 - y_1^2 - 2(0.25 - y_1)y_1 + 0.3^2 - y_2^2 - 2(0.3 - y_2)y_2 + 0.15^2 - y_3^2 - 2(0.15 - y_3)y_3 + 0^2 - y_4^2 - 2(0 - y_4)y_4 + 0.05^2 - y_5^2 - 2(0.05 - y_5)y_5 + 0.^2 - y_6^2 - 2(0 - y_6)y_6 + 0.05^2 - y_7^2 - 2(0.05 - y_7)y_7 + 0.2^2 - y_8^2 - 2(0.2 - y_8)y_8$

$- d_3^R(\vec{\omega}_{\mathcal{K}_3}^E, \vec{y})) = +0.25^2 - y_1^2 - 2(0.25 - y_1)y_1 + 0.13^2 - y_2^2 - 2(0.13 - y_2)y_2 + 0.19^2 - y_3^2 - 2(0.19 - y_3)y_3 + 0^2 - y_4^2 - 2(0 - y_4)y_4 + 0.31^2 - y_5^2 - 2(0.31 - y_5)y_5 + 0.^2 - y_6^2 - 2(0 - y_6)y_6 + 0^2 - y_7^2 - 2(0 - y_7)y_7 + 0.12^2 - y_8^2 - 2(0.12 - y_8)y_8$

$- d_4^R(\vec{\omega}_{\mathcal{K}_4}^E, \vec{y})) = +0.48^2 - y_1^2 - 2(0.48 - y_1)y_1 + 0.12^2 - y_2^2 - 2(0.12 - y_2)y_2 + 0.1^2 - y_3^2 - 2(0.1 - y_3)y_3 + 0^2 - y_4^2 - 2(0 - y_4)y_4 + 0^2 - y_5^2 - 2(0 - y_5)y_5 + 0^2 - y_6^2 - 2(0 - y_6)y_6 + 0.06^2 - y_7^2 - 2(0.06 - y_7)y_7 + 0.24^2 - y_8^2 - 2(0.24 - y_8)y_8$

$- d_5^R(\vec{\omega}_{\mathcal{K}_5}^E, \vec{y})) = +0^2 - y_1^2 - 2(0 - y_1)y_1 + 0.35^2 - y_2^2 - 2(0.35 - y_2)y_2 + 0.34^2 - y_3^2 - 2(0.34 - y_3)y_3 + 0.12^2 - y_4^2 - 2(0.12 - y_4)y_4 + 0.04^2 - y_5^2 - 2(0.04 - y_5)y_5 + 0.12^2 - y_6^2 - 2(0.12 - y_6)y_6 + 0^2 - y_7^2 - 2(0 - y_7)y_7 + 0.05^2 - y_8^2 - 2(0.05 - y_8)y_8$

Using the definition of a probability integrating function (Definition 5.6), we have the probability integrating function $f^R(\vec{y}) = d_1^R(\vec{\omega}_{\mathcal{K}_1}^E, \vec{y})) + d_2^R(\vec{\omega}_{\mathcal{K}_2}^E, \vec{y})) + d_3^R(\vec{\omega}_{\mathcal{K}_3}^E, \vec{y})) +$

Table 5.8 The probability integrating vectors of $\mathcal{R}$ with respect to 2-norm

Vectors	λ	HTD	$HT\bar{D}$	$H\bar{T}D$	$H\bar{T}\bar{D}$	$\bar{H}TD$	$\bar{H}T\bar{D}$	$\bar{H}\bar{T}D$	$\bar{H}\bar{T}\bar{D}$
Γ^A	0.90	0.25	0.15	0.17	0.14	0.06	0.11	0.07	0.05
Γ^C	2.00	0.24	0.15	0.16	0.14	0.09	0.10	0.08	0.05
Γ^{CS}	0.70	0.24	0.16	0.17	0.15	0.04	0.11	0.07	0.05
Γ^{NA}	0.80	0.25	0.15	0.17	0.14	0.06	0.11	0.07	0.05
Γ^{NAI}	0.90	0.25	0.15	0.17	0.14	0.06	0.11	0.07	0.03
Γ^D	0.90	0.26	0.16	0.18	0.15	0.04	0.11	0.07	0.05
Γ^R	2.00	0.25	0.15	0.17	0.14	0.07	0.11	0.07	0.05
Γ^{AI}	1.20	0.28	0.15	0.17	0.14	0.06	0.10	0.06	0.04
Γ^N		0.24	0.15	0.17	0.15	0.06	0.10	0.07	0.05
Γ^E		0.22	0.15	0.14	0.15	0.11	0.09	0.09	0.05
Γ^I		0.26	0.15	0.18	0.14	0.04	0.11	0.06	0.05
Γ^J		0.35	0.03	0.24	0.03	0.01	0.15	0.13	0.06
Γ^L		0.25	0.15	0.17	0.14	0.07	0.11	0.07	0.05
Γ^T		0.35	0.02	0.25	0.02	0.01	0.16	0.13	0.07
Γ^B		0.18	0.20	0.23	0.05	0.14	0.04	0.0024	0.16
Γ^H		0.26	0.13	0.18	0.12	0.04	0.12	0.09	0.05
Γ^Q		0.26	0.13	0.18	0.12	0.04	0.12	0.09	0.05
Γ^K		0.23	0.14	0.15	0.14	0.10	0.10	0.08	0.05
Γ^S		0.25	0.15	0.17	0.14	0.07	0.11	0.07	0.05
Γ^V		0.23	0.19	0.19	0.15	0.04	0.11	0.04	0.05
Γ^M		0.25	0.16	0.18	0.15	0.04	0.11	0.07	0.05
Γ^{HU}		0.28	0.15	0.15	0.10	0.09	0.09	0.09	0.04

Table 5.9 The probability integrating vectors of $\mathcal{R}$ with respect to ∞-norm

Vectors	λ	HTD	$HT\bar{D}$	$H\bar{T}D$	$H\bar{T}\bar{D}$	$\bar{H}TD$	$\bar{H}T\bar{D}$	$\bar{H}\bar{T}D$	$\bar{H}\bar{T}\bar{D}$
Γ^A	0.50	0.25	0.24	0.15	0.03	0.08	0.00	0.06	0.19
Γ^C	2.00	0.24	0.18	0.16	0.12	0.11	0.00	0.06	0.14
Γ^{CS}	0.70	0.27	0.18	0.16	0.09	0.08	0.00	0.07	0.15
Γ^{NA}	0.80	0.26	0.20	0.16	0.08	0.09	0.00	0.06	0.16
Γ^{NAI}	0.90	0.25	0.19	0.16	0.09	0.09	0.00	0.06	0.15
Γ^D	0.90	0.28	0.20	0.17	0.07	0.08	0.00	0.07	0.13
Γ^R	2.00	0.25	0.19	0.16	0.09	0.09	0.00	0.06	0.15
Γ^{AI}	1.20	0.28	0.20	0.16	0.08	0.08	0.00	0.05	0.15
Γ^N		0.28	0.23	0.16	0.00	0.07	0.00	0.07	0.19
Γ^E		0.24	0.16	0.15	0.15	0.11	0.00	0.06	0.12
Γ^I		0.27	0.21	0.16	0.06	0.08	0.00	0.07	0.17
Γ^J		0.22	0.14	0.17	0.12	0.07	0.11	0.07	0.04
Γ^L		0.25	0.19	0.16	0.09	0.09	0.00	0.06	0.15
Γ^T		0.24	0.18	0.11	0.09	0.09	0.00	0.00	0.13
Γ^B		0.25	0.22	0.15	0.04	0.08	0.00	0.06	0.18
Γ^H		0.25	0.22	0.15	0.04	0.08	0.00	0.06	0.18
Γ^Q		0.25	0.22	0.15	0.04	0.08	0.00	0.06	0.18
Γ^K		0.24	0.17	0.16	0.13	0.12	0.00	0.06	0.13
Γ^S		0.25	0.19	0.16	0.09	0.09	0.00	0.06	0.15
Γ^V		0.29	0.19	0.16	0.04	0.07	0.00	0.09	0.17
Γ^M		0.27	0.20	0.16	0.06	0.08	0.00	0.07	0.16
Γ^{HU}		0.30	0.18	0.15	0.05	0.11	0.00	0.05	0.17

$d_4^R(\vec{\omega}_{\mathcal{K}_4}^E, \vec{y})) + d_5^R(\vec{\omega}_{\mathcal{K}_5}^E, \vec{y}))$. *By Theorem 5.8, the probability integrating vector $\vec{y}^{R*}$ of $\mathcal{R}$ is the optimal solution of the following non-linear optimization problem:*

$$\arg \min_{\vec{y} \in \mathbb{P}^8} f^R(\vec{y}) \tag{5.11}$$

subject to (5.8-5.9)

Problem (5.11) has the optimal solotion

$\vec{y}_{\mathcal{B}}^{R*} = (0.25, 0.19, 0.16, 0.09, 0.09, 0.00, 0.06, 0.15)$.

By Definition 5.7, $\Gamma^R(\mathcal{B}) = \vec{y}_{\mathcal{B}}^{R*} = (0.25, 0.19, 0.16, 0.09, 0.09, 0.00, 0.06, 0.15)$. The probability integrating vectors of $\mathcal{R}$ with respect to ∞-norm corresponding to DDFs are shown in Table 5.9.

4) Consider the DDF L-Div. By Definition 5.6 and from Table 5.2, for the un-normalized form, we have the followwing DDFs:

- $d_1^L(\vec{\omega}_{\mathcal{K}_1}^E, \vec{y})) = 0.35 log \frac{0.35}{y_2} + 0.29 log \frac{0.29}{y_3} + 0.06 log \frac{0.06}{y_4} + 0.16 log \frac{0.16}{y_7} + 0.14 log \frac{0.14}{y_8}$

- $d_2^L(\vec{\omega}_{\mathcal{K}_2}^E, \vec{y})) = 0.25 log\frac{0.25}{y_1} + 0.3 log\frac{0.3}{y_2} + 0.15 log\frac{0.15}{y_3} + 0.05 log\frac{0.05}{y_6} + 0.1 log\frac{0.1}{y_7} +$ $0.15\frac{0.15}{y_8}$

- $d_3^L(\vec{\omega}_{\mathcal{K}_3}^E, \vec{y})) = 0.16 log\frac{0.16}{y_1} + 0.22 log\frac{0.22}{y_2} + 0.16 log\frac{0.16}{y_3} + 0.03 log\frac{0.03}{y_4} + 0.31 log\frac{0.31}{y_5} +$ $0.12 log\frac{0.12}{y_7}$

- $d_4^L((\vec{\omega}_{\mathcal{K}_4}^E, \vec{y}))) = 0.2 log\frac{0.2}{y_1} + 0.12 log\frac{0.12}{y_2} + 0.38 log\frac{0.38}{y_3} + 0.06 log\frac{0.06}{y_5} + 0.22 log\frac{0.22}{y_6} +$ $0.02 log\frac{0.02}{y_8}$

- $d_5^L((\vec{\omega}_{\mathcal{K}_5}^E, \vec{y}))) = 0.30 log\frac{0.03}{y_1} + 0.04 log\frac{0.04}{y_2} + 0.46 log\frac{0.46}{y_4} + 0.2 log\frac{0.2}{y_7}$.

Using the definition of a probability integrating function (Definition 5.6), we have the probability integrating function

$$f^L(\vec{y}) = d_1^L(\vec{\omega}_{\mathcal{K}_1}^E, \vec{y})) + d_2^L(\vec{\omega}_{\mathcal{K}_2}^E, \vec{y})) + d_3^L(\vec{\omega}_{\mathcal{K}_3}^E, \vec{y})) + d_4^L(\vec{\omega}_{\mathcal{K}_4}^E, \vec{y})) + d_5^L(\vec{\omega}_{\mathcal{K}_5}^E, \vec{y})).$$

By Theorem 5.8, the probability integrating vector $\vec{y}^{L*}$ of $\mathcal{R}$ is the optimal solution of the following non-linear optimization problem:

$$\arg\min_{\vec{y}\in\mathbb{P}^8} f^L(\vec{y}) \tag{5.12}$$

subject to (5.8-5.9)

Problem (5.12) has the optimal solotion

$\vec{y}_{\mathcal{B}}^{L*} = (0.14, 0.31, 0.19, 0.06, 0.06, 0.01, 0.18, 0.06).$

By Definition 5.7, $\Gamma^L(\mathcal{B}) = \vec{y}_{\mathcal{B}}^{L*} = (0.14, 0.31, 0.19, 0.06, 0.06, 0.01, 0.18, 0.06)$. The probability integrating vectors of $\mathcal{R}$ corresponding to DDFs with respect to the unnormalized form, are shown in Table 5.10.

5) Using the probabilistic integrating operator Γ^{HU} with respect to 1-norm:

- *Finding matrix $\hat{A}_{\mathcal{B}}$*
- *Generating $\vec{\lambda} = (0.1, 0.2, 0.3, 0.3.0.1)$*
- *By Theorem 5.12, $\vec{y}_{\mathcal{B}}^{HU} = \hat{A}_{\mathcal{B}}\vec{\lambda}$*

$$\vec{y}_{\mathcal{B}}^{HU} = \begin{pmatrix} 0 & 0.25 & 0.25 & 0.54 & 0 \\ 0.35 & 0.30 & 0.13 & 0 & 0.34 \\ 0.29 & 0.15 & 0.19 & 0.04 & 0.35 \\ 0.06 & 0 & 0 & 0.12 & 0.11 \\ 0.10 & 0.05 & 0.31 & 0.06 & 0.05 \\ 0 & 0 & 0 & 0 & 0 \\ 0.06 & 0.05 & 0 & 0 & 0.10 \\ 0.14 & 0.20 & 0.12 & 0.24 & 0.05 \end{pmatrix} \begin{pmatrix} 0.1 \\ 0.2 \\ 0.3 \\ 0.3 \\ 0.1 \end{pmatrix} = \begin{pmatrix} 0.29 \\ 0.17 \\ 0.16 \\ 0.05 \\ 0.14 \\ 0 \\ 0.03 \\ 0.17 \end{pmatrix}$$

By Definition 5.8, $\Gamma^{HU}(\mathcal{B}) = \vec{y}_{\mathcal{B}}^{HU} = (0.29, 0.17, 0.16, 0.05, 0.14, 0, 0.030.17)$. The probability integrating vector $\Gamma^{HU}(\mathcal{B})$ of $\mathcal{R}$ with respect to 1-norm are shown in Table 5.7.

Table 5.10 The probability integrating vectors of $\mathcal{R}$ with respect to the unnormalized form

Vectors	λ	HTD	$HT\bar{D}$	$H\bar{T}D$	$H\bar{T}\bar{D}$	$\bar{H}TD$	$\bar{H}T\bar{D}$	$\bar{H}\bar{T}D$	$\bar{H}\bar{T}\bar{D}$
Γ^A	0.50	0.14	0.32	0.19	0.06	0.05	0.01	0.18	0.05
Γ^C	2.00	0.15	0.27	0.17	0.07	0.09	0.02	0.16	0.07
Γ^{CS}	0.70	0.16	0.33	0.20	0.06	0.00	0.00	0.19	0.07
Γ^{NA}	0.80	0.14	0.33	0.20	0.05	0.04	0.01	0.19	0.05
Γ^{NAI}	0.90	0.14	0.32	0.19	0.06	0.05	0.01	0.18	0.05
Γ^D	0.90	0.16	0.36	0.22	0.05	0.00	0.00	0.20	0.00
Γ^R	2.00	0.14	0.31	0.19	0.06	0.06	0.01	0.18	0.06
Γ^{AI}	1.20	0.14	0.35	0.19	0.04	0.05	0.01	0.18	0.04
Γ^N		0.15	0.30	0.18	0.06	0.06	0.01	0.17	0.06
Γ^E		0.15	0.23	0.15	0.08	0.14	0.02	0.14	0.08
Γ^I		0.14	0.35	0.21	0.04	0.01	0.00	0.20	0.04
Γ^J		0.13	0.25	0.15	0.06	0.14	0.03	0.14	0.09
Γ^L		0.14	0.31	0.19	0.06	0.06	0.01	0.18	0.06
Γ^T		0.13	0.25	0.15	0.07	0.15	0.03	0.14	0.09
Γ^B		0.13	0.37	0.22	0.04	0.01	0.00	0.21	0.03
Γ^H		0.13	0.37	0.22	0.04	0.01	0.00	0.21	0.03
Γ^Q		0.13	0.37	0.22	0.04	0.01	0.00	0.21	0.03
Γ^K		0.15	0.25	0.17	0.08	0.11	0.02	0.15	0.07
Γ^S		0.14	0.31	0.19	0.06	0.06	0.01	0.18	0.06
Γ^V		0.20	0.34	0.22	0.04	0.00	0.00	0.20	0.01
Γ^M		0.15	0.35	0.21	0.05	0.00	0.00	0.20	0.05

Step 5: *Calculating the new probability value of probabilistic constraints*

1) Consider the DDF **S**-*Div with respect to* 1-*norm. By Theorem 4.7,*

- For constraint $(H)[\rho_1]$ *then* $\rho_1 = P(HTD) + P(HT\bar{D}) + P(H\bar{T}D) + P(H\bar{T}\bar{D}) = 0.21 + 0.20 + 0.22 + 0.06 = 0.69.$ *Similarly,* $\rho_2 = 0.57$, $\rho_3 = 0.56$.

- For constraint $(T\,|\,H)[\rho_4]$ *then* $\rho_4 = \frac{P(HTD) + P(HT\bar{D})}{\rho_1} = \frac{0.21 + 0.2}{0.69} = 0.59.$ *Similarly,* $\rho_5 = 0.76$.

Therefore, $\mathcal{K}^* = \{(H)[0.69], (T)[0.57], (D)[0.56], (T\,|\,H)[0.59], (H\,|\,D)[0.76]\}$.

After the integration process, by using different DDFs, the new probability value of probabilistic constraints in $\mathcal{K}^*$ *of* $\mathcal{K}$ *with respect to* 1-*norm is shown in Table 5.11.*

Table 5.11 A new PKB $\mathcal{K}^*$ with respect to 1-norm after the integration process when using the DDFs [51]

κ_i	A-Div	C-Div	C-Div	NA-Div	NAI-Div	D-Div	R-Div	AI-Div	N-Div	E-Div	I-Div
(H)	0.67	0.70	0.73	0.69	0.67	0.72	0.69	0.72	0.70	0.70	0.69
(T)	0.56	0.58	0.57	0.56	0.56	0.59	0.57	0.57	0.57	0.58	0.56
(D)	0.52	0.58	0.57	0.56	0.52	0.56	0.56	0.58	0.57	0.59	0.55
$(T\,|\,H)$	0.60	0.60	0.58	0.59	0.60	0.60	0.59	0.60	0.59	0.60	0.59
$(H\,|\,D)$	0.72	0.76	0.83	0.76	0.72	0.78	0.76	0.79	0.78	0.77	0.76
κ_i	J-Div	L-Div	T-Div	B-Div	H-Div	Q-Div	K-Div	S-Div	V-Div	M-Div	HULL-Div
(H)	0.72	0.69	0.64	0.67	0.70	0.70	0.69	0.69	0.77	0.71	0.67
(T)	0.63	0.57	0.51	0.56	0.60	0.60	0.59	0.57	0.51	0.56	0.59
(D)	0.59	0.56	0.58	0.52	0.57	0.57	0.59	0.56	0.63	0.55	0.61
$(T\,|\,H)$	0.65	0.59	0.68	0.60	0.65	0.65	0.60	0.59	0.54	0.58	0.68
$(H\,|\,D)$	0.82	0.76	0.72	0.72	0.77	0.77	0.76	0.76	0.85	0.79	0.74

2) Consider the DDF **B**-*Div with respect to* 1-*norm. Similarly, by Theorem 4.7,* $\mathcal{K}^* = \{(H)[0.67], (T)[0.56], (D)[0.56], (T\,|\,H)[0.57], (H\,|\,D)[0.75]\}$. *After the integration process, by using different DDFs, the new probability value of probabilistic constraints in* $\mathcal{K}^*$ *of* $\mathcal{K}$ *with respect to* 2-*norm shown in Table 5.12.*

Table 5.12 A new PKB $\mathcal{K}^*$ with respect to 2-norm after the integration process when using the DDFs

κ_i	A-Div	C-Div	C-Div	C-Div	NAI-Div	D-Div	R-Div	AI-Div	N-Div	E-Div	I-Div
(H)	0.71	0.69	0.73	0.71	0.71	0.75	0.71	0.75	0.71	0.66	0.73
(T)	0.57	0.57	0.56	0.57	0.57	0.57	0.57	0.59	0.57	0.58	0.57
(D)	0.56	0.56	0.53	0.56	0.56	0.55	0.56	0.57	0.55	0.57	0.55
$(T\,|\,H)$	0.56	0.56	0.56	0.56	0.56	0.56	0.56	0.58	0.56	0.56	0.56
$(H\,|\,D)$	0.76	0.70	0.78	0.77	0.76	0.80	0.75	0.80	0.75	0.65	0.80
κ_i	J-Div	L-Div	T-Div	B-Div	H-Div	Q-Div	K-Div	S-Div	V-Div	M-Div	HULL-Div
(H)	0.64	0.71	0.64	0.67	0.70	0.70	0.67	0.71	0.76	0.73	0.67
(T)	0.54	0.57	0.54	0.55	0.55	0.55	0.58	0.57	0.57	0.56	0.59
(D)	0.73	0.56	0.74	0.56	0.58	0.58	0.57	0.56	0.50	0.54	0.61
$(T\,|\,H)$	0.58	0.56	0.58	0.57	0.56	0.56	0.56	0.56	0.55	0.56	0.68
$(H\,|\,D)$	0.81	0.75	0.82	0.75	0.77	0.77	0.67	0.75	0.84	0.80	0.73

3) Consider the DDF **R**-*Div with respect to* ∞-*norm. Similarly, by Theorem 4.7,* $\mathcal{K}^* = \{(H)[0.69], (T)[0.54], (D)[0.57], (T\,|\,H)[0.64], (H\,|\,D)[0.72]\}$. *After the integra-*

tion process, by using different DDFs, the new probability value of probabilistic constraints in $\mathcal{K}^*$ of $\mathcal{K}$ with respect to ∞-norm shown in Table 5.13.

Table 5.13 A new PKB $\mathcal{K}^*$ with respect to ∞-norm after the integration process when using the DDFs

κ_i	A-Div	C-Div	C-Div	C-Div	NAl-Div	D-Div	R-Div	Al-Div	N-Div	E-Div	I-Div
(H)	0.67	0.70	0.71	0.69	0.69	0.72	0.69	0.72	0.67	0.70	0.69
(T)	0.56	0.53	0.53	0.54	0.54	0.57	0.54	0.56	0.58	0.51	0.55
(D)	0.53	0.57	0.58	0.57	0.57	0.60	0.57	0.57	0.58	0.56	0.57
$(T\,|\,H)$	0.73	0.61	0.64	0.66	0.65	0.67	0.64	0.67	0.76	0.57	0.68
$(H\,|\,D)$	0.74	0.71	0.75	0.73	0.73	0.75	0.72	0.77	0.76	0.69	0.74
κ_i	J-Div	L-Div	T-Div	B-Div	H-Div	Q-Div	K-Div	S-Div	V-Div	M-Div	HULL-Div
(H)	0.65	0.69	0.61	0.68	0.68	0.68	0.70	0.69	0.68	0.70	0.67
(T)	0.53	0.54	0.51	0.56	0.56	0.56	0.52	0.54	0.54	0.55	0.60
(D)	0.52	0.57	0.44	0.55	0.55	0.55	0.56	0.57	0.60	0.58	0.60
$(T\,|\,H)$	0.55	0.64	0.68	0.71	0.71	0.71	0.59	0.64	0.70	0.67	0.67
$(H\,|\,D)$	0.74	0.72	0.78	0.74	0.74	0.74	0.69	0.72	0.74	0.75	0.76

*4) Consider the DDF **L-Div** with respect to the unnormalized form. Similarly, by Theorem 4.7, $\mathcal{K}^* = \{(H)[0.69], (T)[0.52], (D)[0.57], (T\,|\,H)[0.65], (H\,|\,D)[0.58]\}$. After the integration process, by using different DDFs, the new probability value of probabilistic constraints in $\mathcal{K}^*$ of $\mathcal{K}$ with respect to the unnormalized form shown in Table 5.14.*

Table 5.14 A new PKB $\mathcal{K}^*$ with respect to the unnormalized form after the integration process when using the DDFs

κ_i	A-Div	C-Div	C-Div	NA-Div	NAl-Div	D-Div	R-Div	Al-Div	N-Div	E-Div	I-Div
(H)	0.70	0.66	0.74	0.71	0.70	0.79	0.69	0.73	0.69	0.61	0.75
(T)	0.52	0.53	0.49	0.52	0.52	0.53	0.52	0.54	0.52	0.54	0.51
(D)	0.57	0.57	0.54	0.57	0.57	0.58	0.57	0.55	0.57	0.58	0.56
$(T\,|\,H)$	0.65	0.63	0.66	0.65	0.65	0.66	0.65	0.68	0.65	0.62	0.66
$(H\,|\,D)$	0.59	0.55	0.65	0.59	0.59	0.64	0.58	0.59	0.59	0.52	0.62
κ_i	J-Div	L-Div	T-Div	B-Div	H-Div	Q-Div	K-Div	S-Div	V-Div	M-Div	HULL-Div
(H)	0.65	0.69	0.61	0.75	0.75	0.75	0.64	0.69	0.80	0.75	0.67
(T)	0.53	0.52	0.51	0.51	0.51	0.51	0.53	0.52	0.54	0.50	0.58
(D)	0.52	0.57	0.44	0.56	0.56	0.56	0.58	0.57	0.61	0.55	0.61
$(T\,|\,H)$	0.55	0.65	0.68	0.66	0.66	0.66	0.62	0.65	0.67	0.66	0.68
$(H\,|\,D)$	0.74	0.58	0.78	0.61	0.61	0.61	0.54	0.58	0.68	0.65	0.73

5) Using the probability integrating vector Γ^{HU} with respect to 1-norm. Similarly, by Theorem 4.7, $\mathcal{K}^* = \{(H)[0.67], (T)[0.59], (D)[0.61], (T\,|\,H)[0.68], (H\,|\,D)[0.74]\}$. By using the probability integrating vector Γ^{HU}, the integrating results $\Gamma^{HU}(\mathcal{B})$ with respect to 1-norm corresponding to DDFs are shown in Table 5.11, with respect to 2-norm corresponding to DDFs are shown in Table 5.12, with respect to ∞-norm corresponding to DDFs are shown in Table 5.13.

5.5 INTEGRATION ALGORITHMS

5.5.1 Algorithm for Finding the Satisfying Probability Vector

The following algorithm is employed to find a satisfying probability vector of a PKB (**FSPVK**) over a set of events. These vectors are the input of the integration process.

Algorithm FSPVK:

Input: A consistent PKB, a set of events, and pu-norm to identify the problem type

Output: A satisfying probability vector of the PKB over a set of events.

Idea: Depending on pu, the algorithm calls the optimal solution of the corresponding optimization problem. This optimal solution is a satisfying probability vector of the PKB over a set of events.

Algorithm 9 Finding a satisfying probability vector of the PKB (FSPVK)

Input : $\langle \mathcal{K}, \mathsf{E}, pu \rangle$
Output: $\vec{\omega}_{\mathcal{K}}^{\mathsf{E}}$

1 Function FSPVK($\mathcal{K}, \mathsf{E}, pu$)
    begin
2   **if** $pu=1$ or $pu=2$ or $pu=\infty$ **then** $sol = $ NBOP($\mathcal{K}, \mathsf{E}, pu$);
3   **if** $pu=u$ **then** $sol = $ UNOP($\mathcal{K}, \mathsf{E}$);
4      $\vec{\omega}_{\mathcal{K}}^{\mathsf{E}} = $ getOptimalSolution(sol);
5   **return** $\vec{\omega}_{\mathcal{K}}^{\mathsf{E}}$
6 **end**

Algorithm 9 is based on Theorem 5.2, Theorem 5.3, Theorem 5.4, Theorem 5.5, and consists of two stages:

(i) Calling the optimal solution of the corresponding optimization problem with respect to pu (from line 2 to 3).

- If $pu = 1$ or $pu = 2$ or $pu = \infty$, using Algorithm NBOP (line 2).

- If $pu = u$, using Algorithm UNOP (line 3).

(ii) Calling the method to get the optimal solution corresponding to the satisfying probability vector of the input PKB (line 4).

The following theorem evaluates the complexity of Algorithm 9 (**FSPVK**) for finding a satisfying probability vector of the PKB.

Theorem 5.14 *Let $\mathcal{R} = \langle \mathcal{B}, \mathsf{E} \rangle$ be a PKB profile, $\mathcal{K} \in \mathcal{B}$. The complexity of Algorithm 9 (FSPVK) for finding a satisfying probability vector of the PKB is*

   - $\mathcal{O}(\bar{b}_{\mathcal{K}} \times 3^{\bar{b}_{\mathcal{K}}})$ *with respect to 1-norm and ∞-norm*

- $\mathcal{O}(\bar{b}_{\mathcal{K}}^2 \times 3^{\bar{b}_{\mathcal{K}}})$ *with respect to p-norm (p > 1 and $\dot{p} \neq \infty$), and unnormalized form.*

Proof 5.14 *The proof of Theorem 5.14 is very similar to the proof of Theorem 3.16.* □.

5.5.2 The Distance-based Integration Algorithm

The following algorithm is employed to find a probability integrating vector of a PKB profile (**FPIV**) by using the DDFs.

1) Algorithm FPIV:

Input: A PKB profile, the type of the DDF, and *pu*-norm to identify the problem type.

Output: The probability integrating vector of a PKB profile.

Idea: Depending on *pu*, the satisfying probability vector of each PKB in the PKB profile is found. Then, these vectors are combined with the DDFs to build an optimization problem for integrating PKBs. Finally, this optimization problem is solved. The solution of this problem is the probability integrating vector of a PKB profile.

Algorithm 10 Find a probability integrating vector (FPIV) [51]

Input : $d^{\vartheta}, \mathcal{R} = \langle \mathcal{B}, \mathsf{E} \rangle, pu$
Output: $\vec{\varpi}_{\mathcal{R}}$

1 Function FPIV($d^{\vartheta}, \mathcal{R} = \langle \mathcal{B}, \mathsf{E} \rangle, pu$)
   begin
2     $pr \leftarrow$ OptimizationProblem($MINIMIZE$); $Fcc \leftarrow$ FindingSCC(E);
3   sx.setEmpty();
4   **for** $cc \in Fcc$ **do**
5 | pr.addCs($\langle x[cc] \geq 0 \rangle$); sx.addCVar($\langle x[cc] \rangle$);
6   **end for**
7   pr.addCs($\langle sx = 1 \rangle$);
8   f.setEmpty();
9   **for** $\mathcal{K}_i \in \mathcal{B}$ **do**
10        $\vec{\omega}_{\mathcal{K}_i}^{\mathsf{E}} \leftarrow$ FSPVK($\mathcal{K}_i, \mathsf{E}, pu$);
11   **for** $cc \in Fcc$ **do**
12 | $X[i][cc] \leftarrow \vec{\omega}_{\mathcal{K}_i}^{\mathsf{E}}[cc]$;
13 | f.addCVar($\langle$BuildDF($X[i, cc], y[cc], d^{\vartheta}$)$\rangle$);
14   **end for**
15   **end for**
16   pr.addOF($\langle f \rangle$);
17     $solution =$ OpenOpt.solve(NLP, pr);
18     $\vec{\varpi}_{\mathcal{R}} =$ getOptimalSolution($solution$);
19   **return** $\vec{\varpi}_{\mathcal{R}}$;
20 **end**

Algorithm 10 is based on Theorem 5.8 and consists of two stages:

(i) Building the optimization problem (from line 2 to 16), where:

- From line 3 to 7 building constraints $\sum_{j=1}^{\hbar_{\mathsf{E}}} y_j = 1, y_j \geq 0 \forall j = \overline{1, \hbar_{\mathsf{E}}}$.

- From line 8 to 16, building the objective function $f^\vartheta(\vec{y}) = \sum_{i=1}^{\hbar_{\mathcal{B}}} d_i^\vartheta(\vec{\omega}_{\mathcal{K}_i}^{\mathsf{E}}, \vec{y})$, where $X[i][cc]$ is a matrix in which each row is the satisfying probability vectors of $\mathcal{K}_i$ corresponding the complete conjunction cc, $y[cc]$ is a probability integrating vector and d is the form of DDF.

(ii) Solving the optimization problem (5.4) (line 17) with the objective function f (line 16) and getting the optimal value of this problem (line 18).

The following theorem evaluates the complexity of Algorithm 10 (**FPIV**) for finding a probability integrating vector.

Theorem 5.15 ([51]) *Let $\mathcal{R} = \langle \mathcal{B}, \mathsf{E} \rangle$ be PKB profile. The complexity of Algorithm 10 (FPIV) with respect to the DDF d^ϑ:*

- $\mathcal{O}(\hbar_{\mathcal{B}} \times \bar{b}_{\mathcal{K}} \times 3^{\bar{b}_{\mathcal{K}}})$ *with respect to 1-norm and ∞-norm*
- $\mathcal{O}(\hbar_{\mathcal{B}} \times \bar{b}_{\mathcal{K}}^2 \times 3^{\bar{b}_{\mathcal{K}}})$ *with respect to p-norm (p > 1 and $p \neq \infty$), and unnormalized form.*

Proof 5.15 *Consider the DDF S-Div in Table 5.2. The cost of Algorithm 10 depends on the two stages:*

In the first stage, the cost for building the objective function based on the DDF S-Div is $\mathcal{O}(\hbar_{\mathcal{B}} \times 2^{\bar{b}_{\mathcal{K}}})$. For each $\mathcal{K} \in \mathcal{B}$, the cost for finding the satisfying probability vector of $\mathcal{K}$ depends on pu. By Theorem 5.14, the cost is $\mathcal{O}(\bar{b}_{\mathcal{K}} \times 3^{\bar{b}_{\mathcal{K}}})$ with respect to 1-norm and ∞-norm, $\mathcal{O}(\bar{b}_{\mathcal{K}}^2 \times 3^{\bar{b}_{\mathcal{K}}})$ with respect to p-norm (p > 1 and $p \neq \infty$), and unnormalized form. Hence, the cost for finding the satisfying probability vector of $\mathcal{K} \in \mathcal{B}$ is $\mathcal{O}(\hbar_{\mathcal{B}} \times \bar{b}_{\mathcal{K}} \times 3^{\bar{b}_{\mathcal{K}}})$ with respect to 1-norm and ∞-norm, $\mathcal{O}(\hbar_{\mathcal{B}} \times \bar{b}_{\mathcal{K}}^2 \times 3^{\bar{b}_{\mathcal{K}}})$ with respect to p-norm (p > 1 and $p \neq \infty$), and unnormalized form. Therefore, the cost of the first stage is $\mathcal{O}(\hbar_{\mathcal{B}} \times \bar{b}_{\mathcal{K}} \times 3^{\bar{b}_{\mathcal{K}}})$ with respect to 1-norm and ∞-norm, $\mathcal{O}(\hbar_{\mathcal{B}} \times \bar{b}_{\mathcal{K}}^2 \times 3^{\bar{b}_{\mathcal{K}}})$ with respect to p-norm (p > 1 and $p \neq \infty$) and unnormalized form.

In the second stage, the cost for computing $\vec{y}_{\mathcal{B}}^{\mathsf{S}}$ is $\mathcal{O}(N_{\mathsf{S}})$ which depends on the interior-point method [56] to solve Problem (5.4) in Theorem 5.8.

Thus, this cost is $\mathcal{O}\left(max\{\hbar_{\mathcal{B}} \times \hbar_{\mathsf{E}}, N_{\mathsf{S}}\}\right)$. By Problem (3.70) with constraints (3.71-3.72). Moreover, by Problem (5.4),

$$f_0(\vec{y}_{\mathcal{B}}^{\mathsf{S}}) = \sum_{i=1}^{\hbar_{\mathcal{B}}} d^{\mathsf{S}}(\vec{x}_i, \vec{y}) = \sum_{i=1}^{\hbar_{\mathcal{B}}} \sum_{j \in I(y)} (x_{ij} - y_j)^2, \quad Q = (\underbrace{1, \ldots, 1}_{\hbar_{\mathsf{E}}}).$$

By (3.73), $\nabla \Phi(\vec{y}_{\mathcal{B}}^{\mathsf{S}}) = \sum_{i=1}^{\hbar_{\mathsf{E}}} \frac{1}{y_i}$.

Hence, $t_S^{(0)}$ is the minimized value $\inf\limits_{\mathcal{G}} \|t\nabla f_0((\vec{y}_{\mathcal{B}}^{\mathsf{S}})^{(0)}) + \nabla\Phi((\vec{y}_{\mathcal{B}}^{\mathsf{S}})^{(0)}) + Q^T\mathcal{G}\|_2$. By

Problem (3.75) and Problem (4.4), $N_S = \dfrac{\log \frac{\hbar_{\mathsf{E}}}{t_S^{(0)}\varepsilon}}{\log\mu}\left(\dfrac{\hbar_{\mathsf{E}}(\mu-1-\log\mu)}{\gamma}+c\right).$

Let N_S be the cost for computing $\vec{y}_{\mathcal{B}}^{\vartheta}$. Let $g(n) = N_S$ with $n = \bar{b}_{\mathcal{K}}$. Then, $g(n) = \dfrac{n-\log t}{\log(2^{\frac{n}{2}}+1)-\frac{n}{2}}\left(\dfrac{2^{\frac{n}{2}}}{\gamma} - \dfrac{2^n\log(2^{\frac{n}{2}}+1)}{\gamma} + \dfrac{n2^n}{2\gamma} + c\right)$ where t,γ,c are constants.

The fact that $g(n)$ is $\mathcal{O}(n^2 \times 3^n)$. Therefore, the cost for stage 2 is $\mathcal{O}(\bar{b}_{\mathcal{K}}^2 \times 3^{\bar{b}_{\mathcal{K}}})$.

Therefore, the complexity of Algorithm 10 is $\mathcal{O}(\hbar_{\mathcal{B}} \times \bar{b}_{\mathcal{K}} \times 3^{\bar{b}_{\mathcal{K}}})$ with respect to 1-norm and ∞-norm, $\mathcal{O}(\hbar_{\mathcal{B}} \times \bar{b}_{\mathcal{K}}^2 3^{\bar{b}_{\mathcal{K}}})$ with respect to p-norm ($p > 1$ and $p \neq \infty$) and unnormalized form. □.

5.5.3 The HULL Algorithm

The following algorithm employs the operator Γ^{HU} to find a probability integrating vector.

2) Algorithm HULL:

Input: A PKB profile and pu-norm to identify the problem type.

Output: The probability integrating vector of a PKB profile.

Idea: Depending on pu, the satisfying probability vector of each PKB in the PKB profile is found. These vectors are used to build a profile matrix. Then, a vector will be generated such that the elements are in the range $[0,1]$ and its sum is 1. The product of a profile matrix and the vector just found is the probability integrating vector.

Algorithm 11 The probability integrating vector (HULL) [88]

Input : $\mathcal{R} = \langle \mathcal{B}, \mathsf{E} \rangle, pu$
Output: $\vec{\omega}_{\mathcal{R}}$

1 Function HULL($\mathcal{R} = \langle \mathcal{B}, \mathsf{E} \rangle$, pu)
2 **begin**
3   **for** *each* $\mathcal{K}_i \in \mathcal{B}$ **do** $\vec{\omega}_{\mathcal{K}_i}^{\mathsf{E}} \leftarrow$ FSPVK($\mathcal{K}_i, \mathsf{E}, pu$);
4   Finding a profile matrix $\hat{A}_{\mathcal{B}} = \left((\vec{\omega}_{\mathcal{K}_1}^{\mathsf{E}})^T, \ldots, (\vec{\omega}_{\mathcal{K}_{h_{\mathcal{B}}}}^{\mathsf{E}})^T\right)$;
5   Generating vector $\vec{\lambda}$ such that $\lambda_i \in [0,1]$ and $\sum\limits_{i=1}^{\hbar_{\mathcal{B}}} \lambda_i = 1$;
6      $\vec{\omega}_{\mathcal{R}} \leftarrow \hat{A}_{\mathcal{B}}\vec{\lambda}$;
7   **return** $\vec{\omega}_{\mathcal{R}}$;
8 **end**

Algorithm 11 is based on Theorem 5.12. It consists of four stages:

(i) Finding the satisfying probability vector of each PKB in the PKB profile $\mathcal{K} \in \mathcal{B}$ (line 3)

(ii) Finding a profile matrix $\hat{A}_{\mathcal{B}}$. (line 4)

(iii) Generating vector $\vec{\lambda}$ such that $\lambda_i \in [0, 1]$ and $\sum_{i=1}^{\hbar_{\mathcal{B}}} \lambda_i = 1$ (line 5)

(iv) Computing $\hat{A}_{\mathcal{B}}\vec{\lambda}$ (line 6)

The following theorem evaluates the complexity of 11 (**HULL**) for finding a probability integrating vector.

Theorem 5.16 ([88]) *Let $\mathcal{R} = \langle \mathcal{B}, E \rangle$ be PKB profile. The complexity of Algorithm 11 (**HULL**) is:*

- $\mathcal{O}(\hbar_{\mathcal{B}} \times \bar{b}_{\mathcal{K}} \times 3^{\bar{b}_{\mathcal{K}}})$ *with respect to 1-norm and ∞-norm*
- $\mathcal{O}(\hbar_{\mathcal{B}} \times \bar{b}_{\mathcal{K}}^2 \times 3^{\bar{b}_{\mathcal{K}}})$ *with respect to p-norm (p > 1 and p ≠ ∞), and unnormalized form.*

Proof 5.16 *In the first stage, the cost depends on using Algorithm* **FSPVK** *to find the satisfying probability vector of $\mathcal{K}_i \in \mathcal{B}$, denoted by $\bar{N}$. In the second stage, the cost for finding a profile matrix $\hat{A}_{\mathcal{B}}$ is $\mathcal{O}(\hbar_{\mathcal{B}})$. In the third stage, the cost for generating vector $\vec{\lambda}$ is $\mathcal{O}(\hbar_{\mathcal{B}})$. In the final stage, the cost for computing $\hat{A}_{\mathcal{B}}\vec{\lambda}$ is $\mathcal{O}(\hbar_{\mathcal{B}})$. Therefore, the complexity of Algorithm 11 (**HULL**) is $\mathcal{O}(\hbar_{\mathcal{B}}\bar{b}_{\mathcal{K}}3^{\bar{b}_{\mathcal{K}}})$ with respect to 1-norm and ∞-norm, $\mathcal{O}(\hbar_{\mathcal{B}} \times \bar{b}_{\mathcal{K}}^2 \times 3^{\bar{b}_{\mathcal{K}}})$ with respect to p-norm (p > 1 and p ≠ ∞), and unnormalized form.* □

5.6 CONCLUDING REMARKS

This chapter has a brief discussion of the methods for integrating knowledge bases that are needed for understanding and proposing the new method. We then offer a general model for knowledge integration, which are based on the DDFs. The DDFs are essential tools for building the probabilistic integration process. The problems with distance-based integrating the PKBs is analyzed in detail, from theoretic and algorithmic points of view. In order to solve this integration problem, the problems with finding the satisfying probability vector has also been considered and processed. We also offer desired properties of distance-based probabilistic integrating operator which are needed in order to guarantee the chances of obtaining sensible and accurate results. In general, however, the distance-based integration problem is complicated and requires solving the optimization problem. Another technique is discussed later in this chapter.

CHAPTER 6

Value-based method for integrating probabilistic knowledge bases

T HE PURPOSE OF THIS CHAPTER is to deal with the problems of value-based probabilistic knowledge integration. We argue that two particular probabilistic integrating operators which are based on probability values have overall the most appealing properties amongst integrating operators hitherto considered. By investigating these operators, we propose algorithms to practically compute them.

6.1 VALUE-BASED PROBABILISTIC KNOWLEDGE INTEGRATION

6.1.1 Basic Notions

For $c = (F|G)$ and $G \not\equiv \top$, let $\texttt{Left}(c)$ be a function that returns the event on the left of c and $\texttt{Right}(c)$ be a function that returns the event on the right of c, that is, $\texttt{Left}(c) = F$ and $\texttt{Right}(c) = G$. Let $c_1[\rho_1], c_2[\rho_2]$ be probabilistic constraints.

- Two probabilistic constraints $c_1[\rho_1], c_2[\rho_2]$ are called the structural equivalent, denoted by $c_1[\rho_1] \approx c_2[\rho_2]$, iff $\texttt{Left}(c_1) = \texttt{Left}(c_2)$ and $\texttt{Right}(c_1) = \texttt{Right}(c_2)$. Otherwise, probabilistic constraints $c_1[\rho_1], c_2[\rho_2]$ are not called the structural equivalent, denoted by $c_1[\rho_1] \not\approx c_2[\rho_2]$.

- Two probabilistic constraints $c_1[\rho_1], c_2[\rho_2]$ are called the partial equivalent, denoted by $c_1[\rho_1] \simeq c_2[\rho_2]$, iff $\texttt{Left}(c_1) = \texttt{Left}(c_2)$ or $\texttt{Right}(c_1) = \texttt{Right}(c_2)$. Otherwise, two probabilistic constraints $c_1[\rho_1]$ and $c_2[\rho_2]$ are not called the partial equivalent, denoted by $c_1[\rho_1] \not\simeq c_2[\rho_2]$.

There may exist some redundant probabilistic constraints in a PKB. The removal of these probabilistic constraints does not affect the content of the PKB.

Let $\mathbb{C}[\mathcal{Y}] = \mathbb{C}[x_1, x_2, \ldots, x_n]$ be a polynomial ring in n variables. A monomial is an element of the form $x_1^{\alpha_1} \ldots x_n^{\alpha_n}$. A term is an element of the form $cx_1^{\alpha_1} \ldots x_n^{\alpha_n}$ for $c \in \mathbb{C}$ and $\alpha_i \geq 0$.

Definition 6.1 *Let $\mathcal{T} = \{t = x_1^{\alpha_1} \ldots x_n^{\alpha_n} | \alpha_1, \ldots, \alpha_n \in \mathbb{N}_0\}$. A term (or monomial) order $\prec$ is a relation on the monomials of $k[x_1, x_2, \ldots, x_n]$ satisfying the following properties:*

1. $1 \prec t$

2. if $t_1 \prec t_2$ and $t_2 \prec t_3$ then $t_1 \prec t_3$

3. if $t_1 \prec t_2$ then $tt_1 \prec tt_2$

where $t, t_1, t_2, t_3 \in \mathcal{T}$

Definition 6.2 *Let $\mathcal{T} = \{t = x_1^{\alpha_1} \ldots x_n^{\alpha_n} | \alpha_1, \ldots, \alpha_n \in \mathbb{N}_0\}$. A lexicographical term order $\prec_{lex}$ on the set of terms $\mathcal{T}$ is recursively defined by $x_1^{\alpha_1} \ldots x_n^{\alpha_n} \prec_{lex} x_1^{\beta_1} \ldots x_n^{\beta_n}$ iff $\alpha_1 < \beta_1$ or $\alpha_1 = x_2^{\alpha_2} \ldots x_n^{\alpha_n} \prec_{lex} x_2^{\beta_2} \ldots x_n^{\beta_n}$.*

Definition 6.3 *Let $\mathcal{T} = \{t = x_1^{\alpha_1} \ldots x_n^{\alpha_n} | \alpha_1, \ldots, \alpha_n \in \mathbb{N}_0\}$. Let $f = \sum_{i=1}^{n} c_i t_i \in \mathbb{C}[\mathcal{Y}]$ be a polynomial with $c_i \in \mathbb{C} \setminus \{0\}$, $t_i \in \mathcal{T}$, and $t_1 \prec_{lex} t_2 \prec_{lex} \ldots \prec_{lex} t_n$.*

- The leading coefficient of f, denoted by $\mathsf{lc}(f) = c_k$, is the coefficient of the greatest term occurring in f with respect to $\prec_{lex}$.

- The leading monomial of f, denoted by $\mathsf{lm}(f) = t_k$, is the greatest term occurring in f with respect to $\prec_{lex}$.

- The leading term of f is defined by $\mathsf{lt}(f) = \mathsf{lc}(f)\mathsf{lm}(f) = c_k t_k$.

Let $\mathcal{K} = \{(F_1 | G_1)[\rho_1], \ldots, (F_k | G_k)[\rho_k]\}$ be a consistent knowledge base. Taking the conventions $\infty^0 = 1$, $\infty^{-1} = 0$ and $0^0 = 1$ into account.

For each probabilistic constraint $\kappa_i = (F_i | G_i)[\rho_i]$, a normalizing constant α_i of κ_i is defined as follows:

$$\alpha_i = \begin{cases} \infty & \text{if } \rho_i = 1 \\ \in (0, \infty) & \text{if } \rho_i \in (0, 1) \\ 0 & \text{if } \rho_i = 0 \end{cases} \qquad (6.1)$$

and a generalized polynomial f_i of κ_i is defined as follows:

$$
\begin{aligned}
f_i = (1 - x_i)\alpha_i \sum_{\delta(G_iF_i,\Theta)=1} \prod_{\substack{i \neq j \\ \delta(G_jF_j,\Theta)=1}} \alpha_j\alpha_j^{-\rho_j} \prod_{\substack{i \neq j \\ \delta(G_j\overline{F_j},\Theta)=1}} \alpha_j^{-\rho_j} \\
- x_i \sum_{\delta(G_i\overline{F_i},\Theta)=1} \prod_{\substack{i \neq j \\ \delta(G_jF_j,\Theta)=1}} \alpha_j\alpha_j^{-\rho_j} \prod_{\substack{i \neq j \\ \delta(G_j\overline{F_j},\Theta)=1}} \alpha_j^{-\rho_j} = 0
\end{aligned}
\tag{6.2}
$$

Let $\mathcal{SG}(\mathcal{K}) = \{f_1, \ldots, f_n\}$ be a set of generalized polynomials.
Let

$$
f_i^+ = (1 - x_i)\alpha_i \sum_{\delta(G_iF_i,\Theta)=1} \prod_{i \neq j\, \delta(G_jF_j,\Theta)=1} \alpha_j\alpha_j^{-\rho_j} \prod_{i \neq j\, delta(G_j\overline{F_j},\Theta)=1} \alpha_j^{-\rho_j}
\tag{6.3}
$$

$$
f_i^- = -x_i \sum_{\delta(G_i\overline{F_i},\Theta)=1} \prod_{\substack{i \neq j \\ \delta(G_jF_j,\Theta)=1}} \alpha_j\alpha_j^{-\rho_j} \prod_{\substack{i \neq j \\ \delta(G_j\overline{F_j},\Theta)=1}} \alpha_j^{-\rho_j}
\tag{6.4}
$$

$$
f_i^* = \frac{f_i^+ - f_i^-}{\mathsf{gcd}(f_i^+, f_i^-)}
\tag{6.5}
$$

where $\mathsf{gcd}(f_i^+, f_i^-)$ is the greatest common divisor of f_i^+ and f_i^-.

For a further probabilistic constraint $\kappa_{\mathsf{new}} = c_{\mathsf{new}}[\rho_{\mathsf{new}}] = (F_{\mathsf{new}}\,|G_{\mathsf{new}})\,[\rho_{\mathsf{new}}]$, a generalized polynomial f_{new} of κ_{new} is defined as follows:

$$
\begin{aligned}
f_{\mathsf{new}} = x_{\mathsf{new}} \sum_{\delta(G_{\mathsf{new}},\Theta)=1} \prod_{\substack{1 \leq i \leq n \\ \delta(G_iF_i,\Theta)=1}} \alpha_i\alpha_i^{-\rho_i} \prod_{\substack{1 \leq i \leq n \\ \delta(G_i\overline{F_i},\Theta)=1}} \alpha_i^{-\rho_i} \\
- \sum_{\delta(G_{\mathsf{new}}F_{\mathsf{new}},\Theta)=1} \prod_{\substack{1 \leq i \leq n \\ \delta(G_iF_i,\Theta)=1}} \alpha_i\alpha_i^{-\rho_i} \prod_{\substack{1 \leq i \leq n \\ \delta(G_i\overline{F_i},\Theta)=1}} \alpha_i^{-\rho_i} = 0
\end{aligned}
\tag{6.6}
$$

Let

$$
f_{\mathsf{new}}^+ = x_{\mathsf{new}} \sum_{\delta(G_{\mathsf{new}},\Theta)=1} \prod_{\substack{1 \leq i \leq n \\ \delta(G_iF_i,\Theta)=1}} \alpha_i\alpha_i^{-\rho_i} \prod_{\substack{1 \leq i \leq n \\ \delta(G_i\overline{F_i},\Theta)=1}} \alpha_i^{-\rho_i}
\tag{6.7}
$$

$$
f_{\mathsf{new}}^- = \sum_{\delta(G_{\mathsf{new}}F_{\mathsf{new}},\Theta)=1} \prod_{\substack{1 \leq i \leq n \\ \delta(G_iF_i,\Theta)=1}} \alpha_i\alpha_i^{-\rho_i} \prod_{\substack{1 \leq i \leq n \\ \delta(G_i\overline{F_i},\Theta)=1}} \alpha_i^{-\rho_i}
\tag{6.8}
$$

$$f^*_{\text{new}} = \frac{f^+_{\text{new}} - f^-_{\text{new}}}{\gcd(f^+_{\text{new}}, f^-_{\text{new}})} \tag{6.9}$$

Let $\mathcal{SG}^*(\mathcal{K}) = \{f^*_1, \ldots, f^*_n, f^*_{\text{new}}\}$ be a set of further generalized polynomials.

The following definition states that the number of probabilistic constraints, the structure of these constraints and the probability value of these constraints can be changed after reducing some redundant probabilistic constraints in the PKB.

Definition 6.4 ([33, 36, 89]) *(Probabilistic constraint deduction rule) Let* $\mathcal{K} = \{(F_1 | G_1)[\rho_1], \ldots, (F_k | G_k)[\rho_k]\}$. *A PKB* $\mathcal{K}$ *is reduced to* $\mathcal{K}^* = \{(F^*_1 | G^*_1)[\rho^*_1], \ldots, (F^*_h | G^*_h)[\rho^*_h]\}$ *such that* $h \leqslant k$, *denoted by:*

$$\frac{\mathcal{K} : (F_1 | G_1)[\rho_1], \ldots, (F_k | G_k)[\rho_k]}{(F^*_1 | G^*_1)[\rho^*_1], \ldots, (F^*_h | G^*_h)[\rho^*_h]} \tag{6.10}$$

The following theorem states that it could find ρ_{new} from a set of further generalized polynomials and a polynomial ring in n variables.

Theorem 6.1 ([33, 36, 89]) *Let* $\mathcal{K} = \{(F_1 | G_1)[\rho_1], \ldots, (F_k | G_k)[\rho_k]\}$ *be a consistent PKB. Let* $\kappa_{\text{new}} = c_{\text{new}}[\rho_{\text{new}}] = (F_{\text{new}} | G_{\text{new}})[\rho_{\text{new}}]$ *be a further probabilistic constraint. There exists* $(x_1, x_2, \ldots, x_n, x_{\text{new}}) \in \mathbb{C}^{n+1}$ *such that for* $f = \sum_{i=1}^{n} c_i f_i$ *with* $c_i \in \mathbb{C}$ *and* $f_i \in \mathcal{SG}^*(\mathcal{K})$: $\mathcal{SG}^*(\mathcal{K}) \cap \mathbb{C}[x_1, x_2, \ldots, x_n, x_{\text{new}}] = \{f \pm x_{\text{new}}\}$ *implies that* $x_{\text{new}} = f$.

Proof 6.1 *This theorem is proven in full analogy to Theorem 4 [33], so we omit the proofs.* □

The following theorem states that it is possible to eliminate some redundant probabilistic constraint in a PKB by using the Equation (6.11).

Theorem 6.2 ([33, 36, 89]) *Let* $F, G, H \in \mathbf{E}$ *and* $\rho_1, \rho_2 \in \mathbb{R}_{[0,1]}$.

$$\frac{\mathcal{K} : (F | G)[\rho_1], (H | F)[\rho_2]}{(H | G)[\frac{1}{2}(2\rho_1\rho_2 - \rho_1 + 1)]} \tag{6.11}$$

Proof 6.2 *- For* $(F | G)[\rho_1]$,

    + Using formula (6.2), $f_1 = (1 - x_1)\alpha_1(\alpha_2\alpha_2^{-\rho_2} + \alpha_2^{-\rho_2}) - x_1$. *Hence,*

    + By formula (6.3), $f^+_1 = (1 - x_1)\alpha_1(\alpha_2\alpha_2^{-\rho_2} + \alpha_2^{-\rho_2})$

    + By formula (6.4), $f^-_1 = x_1$

    Thus, $\gcd(f^+_1, f^-_1) = 1$

+ By formula (6.5), $f_1^* = (1 - x_1)\alpha_1(\alpha_2\alpha_2^{-\rho_2} + \alpha_2^{-\rho_2}) - x_1$

- For $(H \,|\, F)\,[\rho_2]$, $f_2 = (1 - x_2)\alpha_2\alpha_1\alpha_1^{-\rho_1} - x_2\alpha_1\alpha_1^{-\rho_1}$. Similarly, $f_2^* =$.

Hence,

+ By formula (6.3), $f_2^+ = (1 - x_2)\alpha_2\alpha_1\alpha_1^{-\rho_1}$

+ By formula (6.4), $f_2^- = x_2\alpha_1\alpha_1^{-\rho_1}$

Thus, $\mathbf{gcd}(f_2^+, f_2^-) = \alpha_1\alpha_1^{-\rho_1}$

+ By formula (6.5), $f_2^* = (1 - x_2)\alpha_2 - x_2$

In addition formula (6.6), we have

+ $f_{new} = x_{new}(\alpha_1\alpha_1^{-\rho_1}\alpha_2\alpha_2^{-\rho_2} + \alpha_1\alpha_1^{-\rho_1}\alpha_2^{-\rho_2} + 2\alpha_1^{-\rho_1}) - (\alpha_1\alpha_1^{-\rho_1}\alpha_2\alpha_2^{-\rho_2} + \alpha_1^{-\rho_1})$.

Hence,

+ By formula (6.7), $f_{new}^+ = x_{new}(\alpha_1\alpha_1^{-\rho_1}\alpha_2\alpha_2^{-\rho_2} + \alpha_1\alpha_1^{-\rho_1}\alpha_2^{-\rho_2} + 2\alpha_1^{-\rho_1})$

+ By formula (6.8), $f_{new}^- = \alpha_1\alpha_1^{-\rho_1}\alpha_2\alpha_2^{-\rho_2} + \alpha_1^{-\rho_1}$

Thus, $\mathbf{gcd}(f_{new}^+, f_{new}^-) = \alpha_1^{-\rho_1}$

+ By formula (6.9), $f_{new}^* = x_{new}(\alpha_1\alpha_2\alpha_2^{-\rho_2} + \alpha_1\alpha_2^{-\rho_2} + 2) - (\alpha_1\alpha_2\alpha_2^{-\rho_2} + 1)$.

We have $\mathcal{SG}^*(\mathcal{K}) = \{f_1^*, f_2^*, f_{new}^*\} = \{(1 - x_1)\alpha_1(\alpha_2\alpha_2^{-\rho_2} + \alpha_2^{-\rho_2}) - x_1, (1 - x_2)\alpha_2 - x_2, x_{new}(\alpha_1\alpha_2\alpha_2^{-\rho_2} + \alpha_1\alpha_2^{-\rho_2} + 2) - (\alpha_1\alpha_2\alpha_2^{-\rho_2} + 1)\}$

By Theorem 6.1, we have $\mathcal{SG}^*(\mathcal{K}) \cap \mathbb{C}[x_1, x_2, x_{new}] = \{x_1x_2 - \frac{1}{2}x_1 + \frac{1}{2} - x_{new}\}$ implies that $\rho_{new} = \frac{1}{2}(2\rho_1\rho_2 - \rho_1 + 1)$. □

6.1.2 Value-based Model for Integrating Probabilistic Knowledge Bases

The value-based probabilistic knowledge integration problem is defined as follows:

(1) **Input:** A PKB profile $\mathcal{R} = \langle \mathcal{B}, \mathsf{E} \rangle$.

(2) **Output:** $\mathcal{K}^* = \{(c_1)[\rho_1], \ldots, (c_m)[\rho_m]\}$, $\mathcal{K}^* \not\models \bot$.

(3) **Scope of problem:** The PKBs are represented by probabilistic constraints.

(4) **The integration process:**

 - **Step 1**: Computing an inconsistency measure of the PKBs, it is similar to Step 1 in the distance-based integration model.

 - **Step 2**: Restoring the consistent of the PKBs, it is similar to Step 1 in the distance-based integration model.

 - **Step 3**: The PKBs in $\mathcal{B}$ is integrated by employing Definition 6.6, Definition 6.7 and Theorem 6.2.

A general value-based model for integrating the PKBs is presented in Figure 6.1.

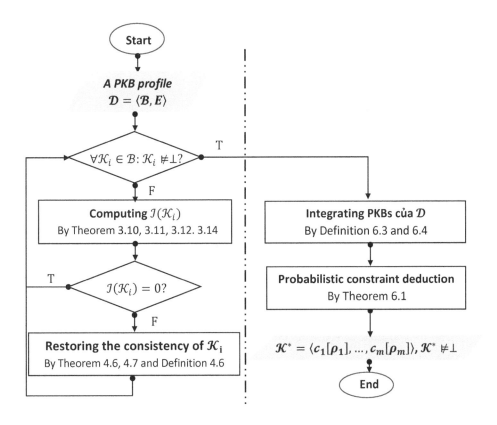

Figure 6.1 The value-based model for integrating probabilistic knowledge bases.

The next section of this chapter will present the theoretical basis for building the algorithms implemented in Step 3 of the integration process.

6.1.3 Desired Properties of Value-based Probabilistic Integrating Operator

In this section, the desirable properties which are used to characterize the value-based probabilistic integrating operators are considered.

Let $\rho^+ = \rho^{1-\rho}$, $\rho^- = \rho^{-\rho}$, $\hat{\rho} = (1-\rho)^{\rho-1}$.

Definition 6.5 ([12, 89]) *Let* $\mathcal{R} = \langle \mathcal{B}, \mathbf{E} \rangle$, $\mathcal{K}_1, \mathcal{K}_2 \in \mathcal{B}$, $(F \,|\, G)\,[\rho_1] \in \mathcal{K}_1$ *and* $(F \,|\, G)\,[\rho_2] \in \mathcal{K}_2$. *An integrating operator* $\bigcirc$ *of two probabilities* ρ_1, ρ_2 *satisfies the following properties:*

(CMT)-Commutativity $\bigcirc(\rho_1, \rho_2) = \bigcirc(\rho_2, \rho_1)$.

*The property **CMT** states that the integrating operator can be executed in any order.*

(IDP)-Idempotence $\bigcirc(\rho_1, \rho_1) = \rho_1$.

*The property **IDP** states that the integrating operator of two equal probabilities does not change.*

(MVP)-Mean Value Property *If $\rho_1 < \rho_2$ then $\rho_1 < \bigcirc(\rho_1, \rho_2) < \rho_2$.*

It implies that the integrating operator of two probabilities is the interval of these probabilities if they are different.

(SFS)-Self-Symmetry $\bigcirc(\rho_1, 1 - \rho_1) = 0.5$.

The self-symmetry means that the integrating operator of two complementary probabilities is equal to 0.5.

(SM)- Symmetry $\bigcirc(1 - \rho_1, 1 - \rho_2) = 1 - \bigcirc(\rho_1, \rho_2)$.

The symmetry means that the integrating operator of the two complement of the probabilities is equal to the complement of the integrating operator of these probabilities.

(SIS)- Semi-Symmetry $\bigcirc(1 - \rho_1, 0.5) = \frac{1 + \rho_1^+ \hat{\rho}_1}{2 + \rho_1^- \hat{\rho}_1}$.

*The property **SIS** states that the integrating operator of a probability's complement and 0.5 can be determined.*

6.2 THE PROBABILITY VALUE-BASED INTEGRATING OPERATORS

In this section, two value-based integrating operators are introduced, that is, the median integrating operator (MIO) and the coefficient median integrating operator (CMIO).

1) The median integrating operator MIO $\oplus$

Based on the integrating operator $\odot$ [12], the following definition presents the median integrating operator of n probabilities corresponding to n structurally equivalent probabilistic constraints, where each probabilistic constraint belongs to a PKB.

Definition 6.6 ([12, 89]) *Let $\mathcal{R} = \langle \mathcal{B}, \mathcal{E} \rangle$, $\mathcal{K}_1, \ldots, \mathcal{K}_n \in \mathcal{B}$; $(F | G) [\rho_1] \in \mathcal{K}_1, \ldots,$ $(F | G) [\rho_n] \in \mathcal{K}_n$. A median integrating operator of n probabilities $\rho_1, \ldots, \rho_n$ $(n \geq 2)$*

is defined as follows:

$$\oplus^n(\rho_1,\ldots,\rho_n) = \frac{\sum\limits_{i=1}^{n} \rho_i^+ \hat{\rho}_i}{\sum\limits_{i=1}^{n} \rho_i^- \hat{\rho}_i} \tag{6.12}$$

Then, the probabilistic constraint $(F\,|G)\,[\oplus^n(\rho_1,\ldots,\rho_n)] \in \mathcal{K}^*$

The following theorem presents the logical relationship between the desirable properties and the median integrating operator.

Theorem 6.3 ([12, 89]) *A median integrating operator* $\oplus^n$ *satisfies* **CMT**, **IDP**, **MVP**, **SFS**, **SM**, *and* **SIS**.

Proof 6.3 *(CMT).* For $n = 2$, we have $\oplus^2(\rho_1,\rho_2) = \frac{\rho_1^+ \hat{\rho}_1 + \rho_2^+ \hat{\rho}_2}{\rho_1^- \hat{\rho}_1 + \rho_2^- \hat{\rho}_2} = \frac{\rho_2^+ \hat{\rho}_2 + \rho_1^+ \hat{\rho}_1}{\rho_2^- \hat{\rho}_2 + \rho_1^- \hat{\rho}_1} = \oplus^2(\rho_2,\rho_1)$, *that is,* $\oplus^n$ *satisfies* **CMT** *for* $n = 2$.
We assume that $\oplus^n$ satisfies **CMT** for $n = k$. It is necessary to show that it follows that $\oplus^n$ satisfies **CMT** for $n = k+1$. Let $\rho = \oplus^k(\rho_1,\ldots,\rho_k)$.
We have $\oplus^2(\rho,\rho_{k+1}) = \frac{\rho^+\hat{\rho} + \rho_{k+1}^+\hat{\rho}_{k+1}}{\rho^-\hat{\rho} + \rho_{k+1}^-\hat{\rho}_{k+1}} = \frac{\rho_{k+1}^+\hat{\rho}_{k+1} + \rho^+\hat{\rho}}{\rho_{k+1}^-\hat{\rho}_{k+1} + \rho^-\hat{\rho}} = \oplus^2(\rho_{k+1},\rho)$, *that is,* $\oplus^n$ *satisfies* **CMT** *for* $n = k+1$.

(IDP). For $n = 2$, we have $\oplus^2(\rho,\rho) = \frac{\rho^+}{\rho^-} = \frac{\rho^{1-\rho}}{\rho^{-\rho}} = \rho$, *that is,* $\oplus^n$ *satisfies* **IDP** *for* $n = 2$. It is easy to see that $\oplus^n(\overbrace{\rho,\ldots,\rho}^{n}) = \rho$.

(MVP). We have $\oplus^2(\rho_1,\rho_2) = \frac{\rho_1^+\hat{\rho}_1 + \rho_2^+\widehat{\rho_2}}{\rho_1^-\hat{\rho}_1 + \rho_2^-\widehat{\rho_2}}$.
As $\rho_1 < \rho_2$, we have $\frac{\rho_1^+\hat{\rho}_1 + \rho_2^+\widehat{\rho_2}}{\rho_1^-\hat{\rho}_1 + \rho_2^-\widehat{\rho_2}} < \frac{2\rho_2^+\widehat{\rho_2}}{2\rho_2^-\widehat{\rho_2}} < \frac{\rho_2^{1-\rho_2}}{\rho_2^{-\rho_2}} = \rho_2$ *and* $\frac{\rho_1^+\hat{\rho}_1 + \rho_2^+\widehat{\rho_2}}{\rho_1^-\hat{\rho}_1 + \rho_2^-\widehat{\rho_2}} > \frac{2\rho_1^+\widehat{\rho_1}}{2\rho_1^-\widehat{\rho_1}} > \frac{\rho_1^{1-\rho_1}}{\rho_1^{-\rho_1}} = \rho_1$. *Hence, this property satisfing* **MVP** *for* $n = 2$. *The proof of operator* $\oplus^n$ *satisfing* **MVP** *is obtained by induction.*

(SFS). It is true for $n = 2$, because we have $\oplus^2(\rho,1-\rho) = \frac{\rho^+\hat{\rho}+(1-\rho)^+\widehat{(1-\rho)}}{\rho^-\hat{\rho}+(1-\rho)^-\widehat{(1-\rho)}} = \frac{\rho(1-\rho)^{-1}+1}{2(1-\rho)^{-1}} = 0.5$. *The proof of operator* $\oplus^n$ *satisfing* **SFS** *is obtained by induction.*

(SM). It is true for $n = 2$, because we have
$\oplus^2(1-\rho_1,1-\rho_2) = \frac{(1-\rho_1)^+\widehat{(1-\rho_1)}+(1-\rho_2)^+\widehat{(1-\rho_2)}}{(1-\rho_1)^-\widehat{(1-\rho_1)}+(1-\rho_2)^-\widehat{(1-\rho_2)}} = \frac{(1-\rho_1)^{\rho_1}\rho_1^{-\rho_1}+(1-\rho_2)^{\rho_2}\rho_2^{-\rho_2}}{(1-\rho_1)^{\rho_1-1}\rho_1^{-\rho_1}+(1-\rho_2)^{\rho_2-1}\rho_2^{-\rho_2}} = 1 - \frac{(1-\rho_1)^{\rho_1-1}\rho_1^{1-\rho_1}+(1-\rho_2)^{\rho_2-1}\rho_2^{1-\rho_2}}{(1-\rho_1)^{\rho_1-1}\rho_1^{-\rho_1}+(1-\rho_2)^{\rho_2-1}\rho_2^{-\rho_2}} = 1 - \oplus^2(\rho_1,\rho_2)$. *The proof of operator* $\oplus^n$ *satisfing* **SM** *is obtained by induction.*

(SIS). It is true for $n = 2$, because we have $\oplus^2(1-\rho_1,0.5) = \frac{\rho_1^+\hat{\rho}_1+0.5^+\widehat{0.5}}{\rho_1^-\hat{\rho}_1+0.5^-\widehat{0.5}} = \frac{(1-\rho_1)^{\rho_1}\rho_1^{-\rho_1}+0.5^{0.5}0.5^{-0.5}}{(1-\rho_1)^{\rho_1-1}\rho_1^{-\rho_1}+0.5^{-0.5}0.5^{-0.5}} = \frac{1+\rho_1^+\hat{\rho}_1}{2+\rho_1^-\hat{\rho}_1}$. *The proof of operator* $\oplus^n$ *satisfing* **SIS** *is obtained by induction.*

The following theorem makes sure that after applying the operator $\oplus^n$, the probabilities are mean values, so that the integrating result reflects average degrees of knowledge as a compromise.

Theorem 6.4 ([12, 89]) $\min\limits_{1\leqslant i\leqslant n}\rho_i \leqslant \oplus^n(\rho_1,\ldots,\rho_n) \leqslant \max\limits_{1\leqslant i\leqslant n}\rho_i$

Proof 6.4 *We have* $\min\limits_{1\leqslant i\leqslant n}\rho_i \leqslant \max\limits_{1\leqslant i\leqslant n}\rho_i$. *According to Theorem 6.3, the operator* $\oplus^n$ *satisfies* **MVP***, so it is easy to get* $\min\limits_{1\leqslant i\leqslant n}\rho_i \leqslant \oplus^n(\rho_1,\ldots,\rho_n) \leqslant \max\limits_{1\leqslant i\leqslant n}\rho_i$.

2) The coefficient median integrating operator CMIO $\ominus_c$

Based on the formula *Bordley* [31], the following definition presents the coefficient median integrating operator of two probabilities corresponding to two structurally equivalent probabilistic constraints, where each probabilistic constraint belongs to a PKB.

Definition 6.7 ([31, 89]) *Let* $\mathcal{R} = \langle\mathcal{B}, \mathcal{E}\rangle$, $\mathcal{K}_1, \mathcal{K}_2 \in \mathcal{B}$, $(F\,|G)\,[\rho_1] \in \mathcal{K}_1$ *and* $(F\,|G)\,[\rho_2] \in \mathcal{K}_2$. *A coefficient median integrating operator of two probabilities* ρ_1, ρ_2 *is defined as follows:*

$$\ominus_c(\rho_1, \rho_2) = \frac{\rho_1^c \rho_2^{1-c}}{\rho_1^c \rho_2^{1-c} + (1-\rho_1)^c (1-\rho_2)^{1-c}} \tag{6.13}$$

where $c \in [0,1]$ *is a coefficient.*

Then, the probabilistic constraint $(F\,|G)\,[\ominus_c(\rho_1, \rho_2)] \in \mathcal{K}^*$.

The following theorem presents the logical relationship between the desirable properties and the coefficient median integrating operator $\ominus_c^2$.

Theorem 6.5 ([31, 89]) *The coefficient median integrating operator* $\ominus_c$ *of two probabilities* ρ_1, ρ_2 *satisfies* **IDP**, **MVP** *and* **SM**.

Proof 6.5 *The proof of Theorem 6.5 is similar to the proof of Theorem 6.3. Hence, we omit the proof.*

The following definition presents the coefficient median integrating operator of n probabilities corresponding to n structurally equivalent probabilistic constraints, where each probabilistic constraint belongs to a PKB.

Definition 6.8 ([31, 89]) *Let* $\mathcal{R} = \langle \mathcal{B}, \mathcal{E} \rangle$, $\mathcal{K}_1, \ldots, \mathcal{K}_n \in \mathcal{B}$; $(F \,|\, G)\,[\rho_1] \in \mathcal{K}_1, \ldots,$ $(F \,|\, G)\,[\rho_n] \in \mathcal{K}_n$. *A coefficient median integrating operator of* n *probabilities* $\rho_1, \ldots, \rho_n$ $(n \geq 2)$ *is defined as follows:*

$$\ominus_c^n(\rho_1, \ldots, \rho_n) = \begin{cases} \dfrac{\rho_1^c \rho_2^{1-c}}{\rho_1^c \rho_2^{1-c} + (1-\rho_1)^c (1-\rho_2)^{1-c}} & \text{if } n = 2. \\ \ominus_c(\ominus_c^{n-1}(\rho_1, \ldots, \rho_{n-1}), \rho_n) & \text{if } n > 2. \end{cases} \tag{6.14}$$

where $c \in [0,1]$ *is a coefficient.*

Then, the probabilistic constraint $(F \,|\, G)\,[\oplus^n(\rho_1, \ldots, \rho_n)] \in \mathcal{K}^*$.

The following theorem presents the logical relationship between the desirable properties and the coefficient median integrating operator $\ominus_c^n$.

Theorem 6.6 ([31, 89]) *The coefficient median integrating operator* $\ominus_c^n$ *of* n *probabilities* **IDP**, **MVP**, *and* **SM**.

Proof 6.6 *The proof of Theorem 6.6 is similar to the proof of Theorem 6.3. Hence, we omit the proof.*

The following theorem makes sure that after applying the operator $\ominus_c^n$, the probabilities are also mean values.

Theorem 6.7 ([31, 89]) $\displaystyle\min_{1 \leqslant i \leqslant n} \rho_i \leqslant \ominus_c^n(\rho_1, \ldots, \rho_n) \leqslant \max_{1 \leqslant i \leqslant n} \rho_i$

Proof 6.7 *We have* $\displaystyle\min_{1 \leqslant i \leqslant n} \rho_i \leqslant \max_{1 \leqslant i \leqslant n} \rho_i$. *According to Theorem 6.3, the operator* $\ominus_c^n$ *satisfies* **MVP**, *so it is easy to get* $\displaystyle\min_{1 \leqslant i \leqslant n} \rho_i \leqslant \ominus_c^n(\rho_1, \ldots, \rho_n) \leqslant \max_{1 \leqslant i \leqslant n} \rho_i$.

6.3   THE PROBABILITY VALUE-BASED INTEGRATION ALGORITHMS

6.3.1   Algorithm for Deducting Probabilistic Constraints

The following algorithm is employed to deduct probabilistic constraints in a PKB (DPC).

Algorithm DPC:

Input: A PKB

Output: A new PKB such that the number of probabilistic constraints in a new PKB is smaller than the one in the original PKB.

Idea: For each pair of two probabilistic constraints, if an event at the right of one constraint equals an event at the left of the other then the reduction rule will be used to derive a probability constraint from the two probabilistic constraints under consideration.

Algorithm 12 Deducting probabilistic constraints (DPC) [89]

Input : $\mathcal{K}$
Output: $\mathcal{K}^*, \bar{b}_{\mathcal{K}^*} < \bar{b}_{\mathcal{K}}$

1 Function DPC($\mathcal{K}$)
    begin
2      $\mathcal{K}^* \leftarrow \mathcal{K}$;
3   **foreach** $\{c_i[p_i], c_j[p_j]\} \subset \mathcal{K}$ *and* $i \neq j$ **do**
4   **if** $\texttt{Left}(c_i) = \texttt{Right}(c_j)$ **then**
5              $c \leftarrow (\texttt{Right}(c_j)\,|\texttt{Left}(c_i)\,)$;
6              $\rho \leftarrow \frac{1}{2}\left(2\rho_i\rho_j - \rho_i + 1\right)$;
7              $\mathcal{K}^* \leftarrow \mathcal{K}^* \cup \{(c)\,[\rho]\}$;
8   **end if**
9   **end foreach**
10   **return** $\mathcal{K}^*$;
11 **end**

Algorithm 12 is based on Theorem 6.2. For each pair of two probabilistic constraints (line 3), if there is an event to the right of the first probabilistic constraint which is similar to one to the left of the second probabilistic constraint (line 4) then:

(i) Replacing these two probabilistic constraints with a new one (from line 5 to 6).

(ii) Adding this new probabilistic constraint to the new PKB $\mathcal{K}^*$ (line 7).

The algorithm 12 terminates when all constrainted pairs in $\mathcal{K}$ are considered.

It is easy to see that if $\mathcal{K} \not\models \perp$ then $DPC(\mathcal{K}) \not\models \perp$.

The following theorem evaluates the complexity of Algorithm 12 (DPC) for deducting probabilistic constraints.

Theorem 6.8 ([31, 89]) *Let $\mathcal{R} = \langle \mathcal{B}, \mathbf{E} \rangle$ be the PKB profile and $\mathcal{K} \in \mathcal{B}$. The complexity of Algorithm 12 (DPC) is*

$$\mathcal{O}\left(\bar{b}_{\mathcal{K}}^2\right) \tag{6.15}$$

Proof 6.8 *Considering the loop in the algorithm, with a pair of two probabilistic constraints in $\mathcal{K}$, in the worst case there will be no pair of two probabilistic constraints that satisfy the condition $\texttt{Left}(c_i) = \texttt{Right}(c_j)$. Hence, the cost depends on the number of probabilistic constraints in $\mathcal{K}$. Therefore, the complexity of the algorithm 12 ((DPC)) is $\mathcal{O}\left(\bar{b}_{\mathcal{K}}^2\right)$.* □

6.3.2 Probability Value-based Integration Algorithms

The following algorithm employs the median integrating operator MIO $\oplus$ and the coefficient median integrating operator CMIO $\ominus_c$ to integrate the PKBs into a consistent PKB.

Algorithms MI-*Mean Integrating*:

Input: A consistent PKB profile, the coefficient median integrating operator op

Output: A consistent PKB.

Idea: Consider two PKBs in the initial PKB profile, check the structural equivalence of two probabilistic constraints belonging to these PKBs. If the two probabilistic constraints are structurally equivalent, the operator op (MIO or $mathttCMIO$) is employed to compute the new probability value of this probabilistic constraint and then add it to the resulting knowledge base. Finally, the redundant probabilistic constraint will be removed from the resulting PKB.

Algorithm 13 The probability value-based integration algorithm (MI) [89]

Input : $\mathcal{R} = \langle \mathcal{B}, \mathsf{E} \rangle, op$
Output: $\mathcal{K}^*$

```
 1  Function MIA(R = ⟨B,E⟩)
    begin
 2      K₀ ← ∅;
 3      foreach Kᵢ ∈ B do
 4          foreach c[ρ] ∈ Kᵢ do
 5              temp ← ρ;
 6              foreach Kⱼ ∈ B and j ≠ i do
 7                  foreach c′[ρ′] ∈ Kⱼ do
 8                      if c′ ≈ c then
 9                          if op = ⊕² then temp ← ⊕²(temp, ρ′) ;
10                          if op = ⊖c then temp ← ⊖c(temp, ρ′) ;
11                          Kⱼ ← Kⱼ − {(c′)[ρ′]};
12                      end if
13                  end foreach
14              end foreach
15          end foreach
16          Kᵢ ← Kᵢ − {(c)[ρ]};
17          K₀ ← K₀ ∪ {(c)[temp]};
18      end foreach
19      K* ← DPC(K₀) ;
20      return K*;
21  end
```

Algorithm 13 is based on Definition 6.6 and Definition 6.7. It consists of two stages:

(i) Finding the common constraints with new probability values (from line 2 to 18)

- Initializing the resulting PKB (line 2).

- Adding the common constraints to the resulting PKB (from line 3 to 18). For each PKB $\mathcal{K}_i \in \mathcal{B}$ (line 3) there exists a probabilistic constraint in $\mathcal{K}_i$ (line 4):

+ Assigning the probability value of the probabilistic constraint under consideration to the probabilistic constraint of the resulting PKB (line 5).

+ Considering the equivalence between a constraint in $\mathcal{K}_i$ with each constraint in the PKBs $\mathcal{K}_j$ (line 6 to 7). If the equivalence is satisfied then:

* Calculating the new probability of the common constraint: line 10 by using the median integrating operator $\oplus$ (Definition 6.6), line 9 by using the coefficient median integrating operator $\ominus_c$ (Definition 6.7).

* Removing the considered constraints from the first PKB in the pair of the PKBs under consideration (line 11)

+ Removing the considered constraints from the second PKB in the pair of the PKBs under consideration (line 16)

+ Adding a new constraint to the resulting PKB (line 17)

(ii) Shortening the redundant probabilistic constraints (line 19) by using Algorithm 12 (DPC).

From Algorithm 13 (MI), it is easy to see that $\mathtt{MI}(\mathcal{B}) \not\models \bot$.

The following theorem evaluates the complexity of Algorithm 13 (MI) for the probability value-based knowledge integration.

Theorem 6.9 ([31, 89]) *Let* $\mathcal{R} = \langle \mathcal{B}, \mathsf{E} \rangle$ *be a PKB profile and* $\mathcal{B} = \{\mathcal{K}_1, \ldots, \mathcal{K}_n\}$. *Let* $m = \max\{\bar{b}_{\mathcal{K}_i} : \forall \mathcal{K}_i \in \mathcal{B}\}$. *The complexity of Algorithm 13 (MI) is*

$$\mathcal{O}\left(\hbar_{\mathcal{B}}^2 \times m^2\right)$$

Proof 6.9 *In the first stage, the estimated cost involves the number of the PKBs in* $\mathcal{B}$ *and the number of constraints of each knowledge base* $\mathcal{K}_i \in \mathcal{B}$:

- Two assignments in line 2 $\mathcal{O}(1)$.

- The cost of adding the common constraints to the resulting PKB (from line 3 to 18):

+ Three assignments in line 5 take $\mathcal{O}(1)$.

+ The number of iterations of loop (line 7) depends on the number of constraints in $\mathcal{K}_j$. *It can be done in* $\mathcal{O}(\bar{b}_{\mathcal{K}_i})$, *where each loop needs:*

* The cost for computing the new probability (line 9) is $\mathcal{O}(1)$;

* The cost for removing the considered constraints from the first PKB in the pair of the PKBs under consideration (line 11) is $\mathcal{O}(1)$;

Thus, the cost for considering the equivalence between two pairs of constraints is $\mathcal{O}(\bar{b}_{\mathcal{K}_i})$.

+ The cost for removing the considered constraints from the second PKB in the pair of the PKBs under consideration (line 16) is $\mathcal{O}(1)$;

+ The cost for adding a new constraint to the resulting PKB (line 17) is $\mathcal{O}(1)$;

Let $m = \max\{\bar{b}_{\mathcal{K}_i} : \forall \mathcal{K}_i \in \mathcal{B}\}$, *we have* $\bar{b}_{\mathcal{K}_i} \leq m \ \forall \mathcal{K}_i \in \mathcal{B}$. *Hence, the cost of the first stage is* $\mathcal{O}\left(\hbar_{\mathcal{B}}^2 \times m^2\right)$.

Let $\mathcal{K}_0$ *be a resulting PKB without reducing probabilistic constraints. We have* $\bar{b}_{\mathcal{K}_0} = \sum_{i=1}^{\hbar_{\mathcal{B}}} \bar{b}_{\mathcal{K}_i} \leq \hbar_{\mathcal{B}} \times m$. *Thus, by Theorem 6.8, the cost of the second stage is* $\mathcal{O}\left(\hbar_{\mathcal{B}}^2 \times m^2\right)$. *Therefore, the complexity of Algorithm 13 is* $\mathcal{O}\left(\hbar_{\mathcal{B}}^2 \times m^2\right)$ □

Example 6.1 *Consider a PKB profile* $\mathcal{R} = \langle \mathcal{B}, \mathsf{E} \rangle$ *in example 2.3, where the consistency of* $\mathcal{K}_1$ *and* $\mathcal{K}_5$ *have been restored with respect to 1-norm as shown in Table 4.5.*

By Definition 3.13, $\mathcal{R}$ *is a consistent PKB profile.*

By using Algorithm 13, the integration process is shown as follows:

- For $(H)[0.7] \in \mathcal{K}_1$: $\mathcal{K}_1 = \{(T)[0.45], (D)[0.45], (T|H)[0.5], (H|D)[0.64]\}$

+ For $(H)[0.7] \in \mathcal{K}_2$, $(H)[0.7] \approx (H)[0.7]$, *by Definition 6.6*

$temp = \oplus(0.7, 0.7) = \frac{0.7^{0.3}0.3^{-0.3} + 0.7^{0.3}0.3^{-0.3}}{0.7^{-0.7}0.3^{-0.3} + 0.7^{-0.7}0.3^{-0.3}} = 0.7$ *and*

$\mathcal{K}_2 = \{(T)[0.6], (D)[0.5], (T|H)[0.78], (H|D)[0.8]]\}$

+ For $(H)[0.57] \in \mathcal{K}_3$, $(H)[0.7] \approx (H)[0.57]$,

$temp = \oplus(0.7, 0.57) = \frac{0.7^{0.3}0.3^{-0.3} + 0.57^{0.43}0.43^{-0.43}}{0.7^{-0.7}0.3^{-0.3} + 0.57^{-0.57}0.43^{-0.43}} = 0.63$ *and*

$\mathcal{K}_3 = \{(T)[0.69], (D)[0.75], (T|H)[0.67]\}$

+ For $(H)[0.7] \in \mathcal{K}_4$, $(H)[0.7] \approx (H)[0.7]$, $temp = \oplus(0, 63, 0.7) = 0.67$ *and* $\mathcal{K}_4 = \{(T)[0.6], (D)[0.64], (H|D)[0.9]\}$

+ For $(H)[0.8] \in \mathcal{K}_5$, $(H)[0.67] \approx (H)[0.7]$, $temp = \oplus(0, 67, 0.8) = 0.73$ *and* $\mathcal{K}_5 = \{(T)[0.38], (D)[0.5], (T|H)[0.42], (H|D)[0.71]\}$

Hence, $\mathcal{K}_0 = \{(H)[0.73]\}$

- For $(T)[0.3] \in \mathcal{K}_1$, *similarly* $temp = 0.49$ *and*

$\mathcal{K}_1 = \{(D)[0.45], (T|H)[0.5], (H|D)[0.64]\}$, $\mathcal{K}_3 = \{(D)[0.75], (T|H)[0.67]\}$

$\mathcal{K}_2 = \{(D)[0.5], (T|H)[0.78], (H|D)[0.8]]\}$, $\mathcal{K}_4 = \{(D)[0.64], (H|D)[0.9]\}$,

$\mathcal{K}_5 = \{(D)[0.5], (T|H)[0.42], (H|D)[0.71]\}$

Hence, $\mathcal{K}_0 = \{(H)[0.73], (T)[0.49]\}$

- For $(D)[0.45] \in \mathcal{K}_1$, *similarly* $temp = 0.56$ *and*

$\mathcal{K}_1 = \{(T|H)[0.5], (H|D)[0.64]\}$, $\mathcal{K}_2 = \{(T|H)[0.78], (H|D)[0.8]]\}$,

$\mathcal{K}_3 = \{(T|H)[0.67]\}$, $\mathcal{K}_4 = \{(H|D)[0.9]\}$, $\mathcal{K}_5 = \{(T|H)[0.42], (H|D)[0.71]\}$

Hence, $\mathcal{K}_0 = \{(H)[0.73], (T)[0.49], (D)[0.56]\}$

- For $(T|H)[0.5] \in \mathcal{K}_1$, *similarly temp* $= 0.54$ *and*

$\mathcal{K}_1 = \{(H|D)[0.64]\}$, $\mathcal{K}_2 = \{(H|D)[0.8]]\}$,

$\mathcal{K}_3 = \varnothing$, $\mathcal{K}_4 = \{(H|D)[0.9]\}$, $\mathcal{K}_5 = \{(H|D)[0.71]\}$

Hence, $\mathcal{K}_0 = \{(H)[0.73], (T)[0.49], (D)[0.56], (T|H)[0.54]\}$

- For $(H|D)[0.64] \in \mathcal{K}_1$, *similarly temp* $= 0.69$ *and* $\mathcal{K}_1 = \mathcal{K}_2 = \mathcal{K}_4 = \mathcal{K}_5 = \varnothing$.

Therefore, $\mathcal{K}_0 = \{(H)[0.73], (T)[0.49], (D)[0.56], (T|H)[0.54], (H|D)[0.69]]\}$.

After using the median integrating operator and the coefficient median integrating operator, the resulting PKB $\mathcal{K}_0$ *is shown in Table 6.1.*

Table 6.1 The resulting PKB $\mathcal{K}_0$ after using the integrating operators $\oplus$ MIO and CMIO

κ_i	$\mathtt{MI}(\mathcal{R}, \oplus^2)$	$\mathtt{MI}(\mathcal{R}, \ominus_0)$	$\mathtt{MI}(\mathcal{R}, \ominus_{0.3})$	$\mathtt{MI}(\mathcal{R}, \ominus_{0.5})$	$\mathtt{MI}(\mathcal{R}, \ominus_{0.7})$	$\mathtt{MI}(\mathcal{R}, \ominus_1)$
(H)	0.73	0.80	0.77	0.74	0.72	0.70
(T)	0.49	0.38	0.45	0.50	0.52	0.50
(D)	0.56	0.50	0.55	0.57	0.56	0.45
$(T|H)$	0.54	0.42	0.51	0.55	0.56	0.50
$(H|D)$	0.69	0.60	0.70	0.73	0.73	0.64

By Algorithm 12,

$\mathcal{K}^* = \mathtt{DPC}(\mathcal{K}^*) = \{(H)[0.73], (T)[0.49], (D)[0.56], (T|D)[0.53]\}$

After applying Algorithm 12 (DPC) to reduce the probabilistic constraints, the PKBs are shown in Table 6.2. It is easy to see that the smaller the value of c is, the larger the value of probability is. The probability values always fulfill Theorem 6.4 and Theorem 6.7.

Table 6.2 The PKB $\mathcal{K}^*$ after reducing constraints

κ_i	$\mathtt{MI}(\mathcal{R}, \oplus^2)$	$\mathtt{MI}(\mathcal{R}, \ominus_0)$	$\mathtt{MI}(\mathcal{R}, \ominus_{0.3})$	$\mathtt{MI}(\mathcal{R}, \ominus_{0.5})$	$\mathtt{MI}(\mathcal{R}, \ominus_{0.7})$	$\mathtt{MI}(\mathcal{R}, \ominus_1)$
(H)	0.73	0.80	0.77	0.74	0.72	0.70
(T)	0.49	0.38	0.45	0.50	0.52	0.50
(D)	0.56	0.50	0.55	0.57	0.56	0.45
$(T|D)$	0.53	0.45	0.50	0.54	0.55	0.50

6.4 CONCLUDING REMARKS

In this chapter, a value-based model for integrating the PKBs is presented. We introduced two probability value-based integrating operators (MIO, CMIO) and showed that all members satisfy many desired properties that have been considered in the literature. In order to eliminate the redundant probabilistic constraint in the resulting knowledge base, the probabilistic constraint reduction rule is employed so that the reduced knowledge base is still equivalent to the original knowledge base.

CHAPTER 7

Experiments and applications

T HE GOAL OF THIS CHAPTER is to show our experimental data along with analysis and to address the difficulties raised previously.

7.1 EXPERIMENT

7.1.1 Experimental Purpose and Assumptions

The main objectives to conduct these experiments are as follows:

(i) Evaluate the quality of the integrating results by ultilizing different DDFs and norms. Thereby, it confirms that utilizing DDFs for integrating process will lead to the results that adheres to probabilistic principles.

(ii) Examine the impact of coefficient selection in DDFs on integrating outcomes.

(iii) Provide evaluations and comparisons of the number of iterations, CPU time, and overall time spent to implement the integrating process utilizing various DDFs and norms. As a result, the performance of the integrating algorithms is reported to be feasible and executable.

(iv) Evaluate the performance of integrating process using various input size test sets. Based on this analysis, it justifies that the input size has an impact on the integrating process's efficiency.

The empirical estimates of the experiments are as follows:

(i) Prediction of integrating results: A probability integrating vector x^* of $\mathcal{R}$ is the optimal solution of the non-linear optimization problem (5.4) that must satisfy condition $\sum_{i=1}^{\hbar_E} x_i = 1, x_i \geq 0$ based on the constraint set C_m. As a result, the value of probabilistic constraints in integrating PKBs will meet the basic principle of probability $0 \leq \mathcal{P}(A) \leq 1$ for all $A \in \mathbf{E}$, as follows from Theorem 2.13 [37]: $P(A) = \sum_{i=1}^{k} P(B_i)P(A|B_i)$, hence $0 \leq P(A|B_i) \leq 1$.

(ii) Prediction of integrating results in DDFs with various coefficients: According to Table 5.3, coefficient λ in A-Div and NA-Div must satisfy condition $\lambda \neq 1$, whereas coefficient λ in CS-Div, R-Div, and others must satisfy conditions $\frac{-\Pi}{2} \leq \lambda \leq \frac{\Pi}{2}$, $\lambda \in \mathbb{R}_{(1,2]}$, and $\lambda \neq 1$. As a result, the coefficients' values have a significant impact on the integrating results. Thus, the value of coefficients within the constrained domain can be stated to produce superior integrating results.

(iii) Cost estimation for solving integrating algorithms: Finding realistic estimates for the performance of integrating PKBs is highly dependent on the algorithm used and the PKBs' structure. Some empirical estimations for integrating algorithms are described in Table 7.1 based on five constraint sets C_1, C_2, C_∞, C_r and C_m relating to various DDFs and norms. Matousek and Gartner [71] estimate $2m$ to $3m$ iterations when using the Simplex method for linear optimization problems originating in Theorem 5.3, and Theorem 5.4, where m is the number of constraints after converting the problem to a standard form including equations and non-negative constraints. The number of events in $\mathbf{E}$ is $\hbar_\mathsf{E}$. Each iteration often results in costs in the order of $2^{|\mathsf{E}|}|\mathcal{K}|$, resulting in an overall cost in the order of $2^{\hbar_\mathsf{E}}\bar{b}_\mathcal{K}$. The interior-point methods presented in [56] can solve the non-linear optimization problems arising in Theorem 4.6, Theorem 5.2, and Theorem 5.8 in around 10 to 100 iterations with cost $(2^{\hbar_\mathsf{E}})^2\bar{b}_\mathcal{K}$ per iteration, according to practical studies.

Table 7.1 The number of optimization variables n, The number of constraints m, Estimate [51]

Algorithms	Norm	n	m	Estimate
FSPVK	1-norm	$2^{\hbar_\mathsf{E}} + \bar{b}_\mathcal{K}$	$2\bar{b}_\mathcal{K} + 3$	$n\,m^2$
FSPVK	2-norm	$2^{\hbar_\mathsf{E}}$	2	$n^2\,m$
FSPVK	∞-norm	$2^{\hbar_\mathsf{E}} + 1$	$2\bar{b}_\mathcal{K} + 3$	$n\,m^2$
RCK		$2^{\hbar_\mathsf{E}} + \bar{b}_\mathcal{K}$	$\bar{b}_\mathcal{K} + 4$	$n^2\,m$
FPMV		$2^{\hbar_\mathsf{E}}$	2	$n^2\,m$

Overall, the experimental results show that when utilizing DDFs, the proposed strategy of integrating PKBs produces the expected effects for three norms.

7.1.2 Experiment Settings

Integrating Algorithms. We ran comprehensive tests to evaluate the performance of integrating PKBs from a PKB profile by utilizing integrating algorithms based on norms (1-norm, 2-norm, ∞-norm) and DDFs.

Environment. All of the proposed algorithms were implemented in Java (`https://www.oracle.com/java`) and Python (`https://www.python.org/`). The following tests were performed on a PC running 64-bit Windows Server 2012R2, an Intel (R) Xeon(R) E5-2686 v4 @2.3GHz CPU, and 16 GB RAM.

Dataset. The Cancer Symptoms Survey was released in 2012 and can be found at `https://www.quality-health.co.uk`. It shows the incidence of cancers and the association between cancers and a group of given symptoms. We use the results of this survey to create a list of events, which includes:

- Three cancer types are considered, namely Chronic Myeloid Leukemia (M), Chronic Lymphocytic Leukemia (C), and Lung Cancer (L).

- Five related symptoms: Extreme fatigue/tiredness (denoted by T), Unusual sweating at night (denoted by N), Unexpected weight loss (denoted by W), Shortness of breath (denoted by B), Lump in neck/groin/armpit (denoted by A).

The relationship between three types of cancer and five associated symptoms for experimental data are presented in Table 7.2.

Table 7.2 The set of events depicts the links among three cancer types and five symptoms [51]

Related Symptoms	M	C	L
	0.77	0.74	0.67
Extreme fatigue/tiredness(T)	0.51	0.33	0.17
Unusual sweating at night(N)	0.33	0.24	-
Unexpected weight loss(W)	0.27	-	-
Shortness of breath(B)	0.19	0.23	0.31
Lump in neck/groin/armpit(A)	-	0.35	-

When a person has Chronic Myeloid Leukemia, there is a 0.65 chance that they may develop Chronic Lymphocytic Leukemia. Since then, the following set of probabilistic constraints has been discovered: $(M)[0.77]$, $(C)[0.74]$, $(L)[0.67]$, $(T|M)[0.51]$, $(T|C)[0.33]$, $(N|M)[0.33]$, $(N|C)[0.24]$, $(W|M)[0.27]$, $(B|M)[0.19]$,

$(B|C)[0.23]$, $(B|L)[0.31]$, $(A|C)[0.35]$, $(T|L)[0.17]$, $(C|M)[0.65]$. Test sets with PKB profiles encompassing nine PKBs are constructed based on a set of these limitations. The number of events and probabilistic constraints are increased in order to detect the sensitivity of various integrating techniques to the problem size. Table 7.3 provides more information about the test sets.

Table 7.3 Test sets [51]

The problem size	Test 1	Test 2
The number of knowledge bases	9	9
The number of events	5	8
The number of complete conjunctions	32	256
The number of probabilistic constraints	8	14

The PKB profile of Test 1 is $\mathcal{D} = \langle \mathcal{B}, \mathsf{E} \rangle$, where $\mathsf{E} = \{L, M, C, T, A\}$ and $\mathcal{B}$ including nine PKBs with the probabilistic constraints are described in Table 7.4.

Table 7.4 PKBs for Test 1 [51]

Constraints	$\mathcal{K}_1$	$\mathcal{K}_2$	$\mathcal{K}_3$	$\mathcal{K}_4$	$\mathcal{K}_5$	$\mathcal{K}_6$	$\mathcal{K}_7$	$\mathcal{K}_8$	$\mathcal{K}_9$	Min	Mean Value	Max
(L)	0.65	0.62	0.64	0.62	0.65	0.61	0.61	0.60	0.62	0.60	0.62	0.65
(M)	0.79	0.71	0.71	0.71	0.79	0.72	0.72	0.70	0.81	0.70	0.73	0.81
(C)	0.70	0.78	0.76	0.80	0.70	0.70	0.70	0.73	0.72	0.70	0.74	0.8
$(T|M)$	0.50		0.45	0.54	0.50	0.57	0.57	-	-	0.45	0.53	0.57
$(T|C)$	-	0.32	0.19	0.36	0.30	0.39	0.39	0.50	0.44	0.19	0.36	0.5
$(A|C)$	-	0.34	0.32	0.37	-	0.31	0.31	0.30	0.32	0.30	0.32	0.37
$(T|L)$	0.16	0.18	0.19	-	0.16	0.15	0.12	0.18	0.19	0.12	0.17	0.19
$(C|M)$	0.65	-	-	-	-	-	0.60	0.70	0.70	0.60	0.67	0.7

The PKB profile of Test 2 is $\mathcal{D} = \langle \mathcal{B}, \mathsf{E} \rangle$, where $\mathsf{E} = \{L, M, N, A, B, C, T, W\}$ and $\mathcal{B}$ including nine PKBs with the probabilistic constraints are described in Table 7.5.

Table 7.5 PKBs for Test 2 [51]

Constraints	$\mathcal{K}_1$	$\mathcal{K}_2$	$\mathcal{K}_3$	$\mathcal{K}_4$	$\mathcal{K}_5$	$\mathcal{K}_6$	$\mathcal{K}_7$	$\mathcal{K}_8$	$\mathcal{K}_9$	Min	Mean Value	Max
(L)	0.67	0.67	0.69	0.68	0.7	0.6	0.62	0.6	0.66	0.60	0.65	0.7
(M)	0.77	0.75	0.7	0.72	0.8	0.75	0.73	0.7	0.71	0.70	0.73	0.8
(C)	0.74	0.73	0.72	0.71	0.72	0.77	0.75	0.8	0.81	0.71	0.75	0.81
$(T\,|\,M)$	0.51	0.52	-	0.54	0.45	0.56	0.55	0.53	0.57	0.45	0.53	0.57
$(T\,|\,C)$	0.33	0.3	-	0.29	0.36	0.32	0.36	0.29	0.31	0.29	0.32	0.36
$(N\,|\,M)$	0.33	0.33	0.3	-	0.3	0.29	0.3	0.31	-	0.29	0.31	0.33
$(N\,|\,C)$	0.24	0.21	0.22	-	0.21	0.21	0.22	0.24	0.3	0.21	0.23	0.3
$(W\,|\,M)$	0.27	0.27	0.25	0.21	-	0.3	0.29	-	0.2	0.20	0.25	0.3
$(B\,|\,M)$	0.19	-	0.15	0.2	0.2	0.15	0.21	-	0.23	0.15	0.19	0.23
$(B\,|\,C)$	0.23	-	0.29	0.21	0.25	0.26	0.19	0.18	0.26	0.18	0.23	0.29
$(A\,|\,C)$	0.35	0.36	0.32	0.37	0.36	-	-	0.31	0.34	0.31	0.34	0.37
$(T\,|\,L)$	0.17	0.18	-	0.15	0.13	0.18	0.18	0.21	-	0.13	0.17	0.21
$(B\,|\,L)$	0.31	-	0.34	0.33	0.32	0.35	0.32	0.37	0.32	0.31	0.34	0.37
$(C\,|\,M)$	0.7	-	0.65	-	0.73	-	-	0.7	-	0.65	0.69	0.73

7.1.3 Experimental Implementation

PKB must be represented and optimization problem must be solved in the integration process. In order to develop libraries to represent PKBs, several libraries from the Tweety-1.14 project (`http://mthimm.de/projects/tweety/`) are utilized and inherited.

- Finding a satisfying probability vector and an inconsistency measure with respect to 1-norm and ∞-norm necessitates solving linear programming problems with optimal solutions satisfying sets C_1 and C_∞ respectively.

- The algorithm for determining the satisfying probability vector and the inconsistency measure with respect to the 2-norm says that it can solve quadratic problems with the best solution satisfying set C_2..

- Finding the probability integrating vector necessitates solving non-linear optimization problems with the optimal solution satisfying set C_m.

- The algorithm for restoring consistency claims to solve a non-linear optimization issue by finding the optimal solution that meets the set C_r.

The OpenOpt optimization library (`https://pypi.org/project/openopt/`) developed in Python has been used to solve two important categories of optimization problems including linear programming problem, non-linear optimization problem by installing OpenOpt 0.5629 on our system (`https://packaging.python.org/tutorials/installing-packages/`).

In OpenOpt, there are many solvers to solve the class of optimization problem such as Least Squares Problem (LSP), Linear Least Squares Problem (LLSP), Linear Problem (LP), Non-Linear Problem (NLP), Non-Linear Solve Problem (NLSP),

Mixed-Integer Linear Problem (MILP), Quadratic Problem (QP), Non-Smooth Problem (NSP), Mini-Max Problem (MMP), and Global Problem (GLP).

They can be tackled with OpenOpt by configuring Linear Problem (LP) in the system by establishing the Apache Commons Simplex solver (`http://commons.apache.org/math`) before launching reasoning task for the first time.

Non-Linear Problem (NLP) is a type of non-linear optimization problem that may be solved using OpenOpt by specifying it in the system.

OjAlgo (`https://www.ojalgo.org/`), an Open Source Java code for mathematics, linear algebra, and optimization, can be used to solve quadratic programming problems. Java was also used to implement the inconsistency measures, discover the satisfying probability vectors, determine the satisfying restored probability vector, find the probability integrating vector, and compute the new probability value. The PKBSPROJECT project produces all integration process algorithms.

The source code of this project is stored in a JAR file that may be obtained in an online appendix (`https://thamnguyenvan.blogspot.com/2020/04/explaination.html`). It is important to utilize Eclipse (`https://www.eclipse.org`) to install the PKBSPROJECT libraries.

7.1.4 Results and Analysis

This section will assess and analyze the outcomes obtained using the integrating model described in Figure 5.2 in three ways:

- The quality of the Test 1 and Test 2 outcomes acquired in terms of three norms (1-norm, 2-norm, and ∞-norm).

- The impact of coefficient selection in DDFs on the final results.

- The cost of implementing the recommended algorithms by comparing the Test 1 and Test 2 outcomes.

An online appendix (`https://thamnguyenvan.blogspot.com/`) contains details of the achieved outcomes.

a) Integrating quality

The quality of the integrating results will be evaluated in this stage. The probability value of each constraint in the final PKB that results from the integration process determines this quality. For each probabilistic constraint $c^*[\rho]$, the resulting $\mathcal{K}$ will consider whether ρ is in the interval $[a, b]$ or not; where a and b are the minimum and maximum probability values of $c_i[a] \in \mathcal{K}_i$ and $c_j[b] \in \mathcal{K}_j$, respectively, such that c^*, c_i, c_j are structurally comparable.

However, before we go into the details, it is important to clarify what is expected from the result as constraints by a very basic Theorem 5.1. This theorem assures that the integrating probabilities are close to mean values, resulting in a probability of restrictions that represents average value of integrating degrees. The results of Tests

1 and 2 are used to assess the quality of the experimental results produced. For each constraint, the minimal, expected (average), and maximum probability values are listed in Table 7.4 for Test 1 and Table 7.5 for Test 2.

The criteria for evaluating the obtained results
- The results of Test 1 will be shown in Figures 7.1.
- The results of Test 2 will be shown in Figures 7.2.

It indicates that the probability value (y-axis) of constraints (x-axis) in the merged knowledge base satisfied three conditions after employing integrating algorithms based on DDFs with respect to three norms (1-norm, ∞-norm, 2-norm) as follows:

(i) By Definition 2.9 [37], the fundamental principle of probability is $0 \leq \mathcal{P}(A) \leq 1$;

(ii) By Definition 2.10 [37], the conditional probability of B, given A, $\mathcal{P}(A|B) = \frac{\mathcal{P}(AB)}{\mathcal{P}(B)}$ supplied $\mathcal{P}(B) > 0$.

(iii) Principle of expectation by Theorem 5.1.

The dark red line with red points in Figure 7.1 and Figure 7.2 shows the expected probability values of probabilistic constraints that should be attained, while the remaining dotted line with points reflects the probability value of the constraints obtained for each DDF employed.

The dots on dotted lines are very close to the dot on the dark red line. Furthermore, dotted line points range from 0 to 1. As a result, it is concluded that the achieved outcomes are close to those desired. In the three sub-figures, several points of results are slightly different from those expected, indicating that there are some biases (Δ). This bias, however, is acceptable according to Theorem 5.1 for two reasons.

Firstly, this bias is really minor. As a result, the obtained values are rather near to the desirable values.

Secondly, the joint PKB is assured to be consistent. For Test 1, After employing the DDF **A**-Div and performing with respect to 1-norm in the integration stage, the resulting PKB $\mathcal{K}^*$ of nine input PKBs is a desirable result. As a result, $\Delta = 0$ for the constraint (M) and $\Delta = 0.06$ for the constraint $(T|L)$ satisfy Theorem 5.1. Furthermore, $\mathcal{I}(\mathcal{K}) = 0$.

Overall, the acquired PKBs are consistent with the original forecast of the quality of the integration stage employed DDFs that adhere to a predetermined standard. The quality of the probabilistic constraint for both test sets is acceptable in terms of three norms, as shown in Figure 2.1.

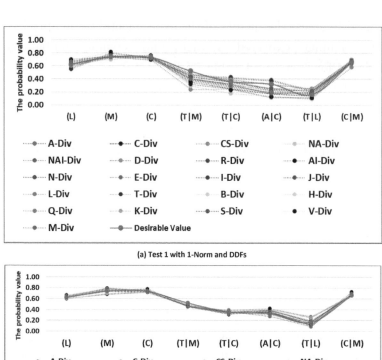

(a) Test 1 with 1-Norm and DDFs

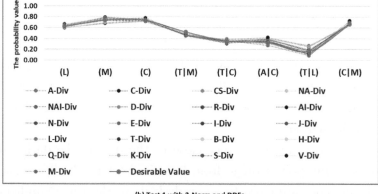

(b) Test 1 with 2-Norm and DDFs

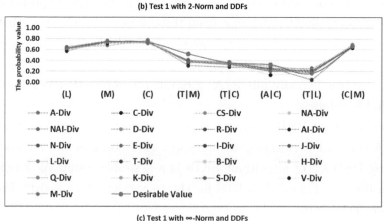

(c) Test 1 with ∞-Norm and DDFs

Figure 7.1 Comparison of the quality of probabilistic constraints after the integration process with Test 1 [51].

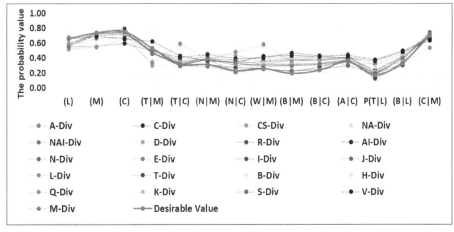

(a) Test 2 with 1-Norm and DDFs

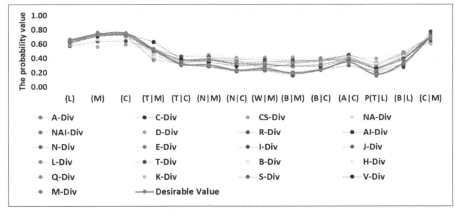

(b) Test 2 with 2-Norm and DDFs

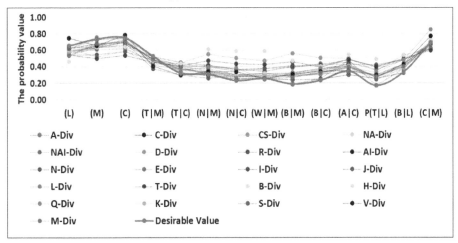

(c) Test 2 with ∞-Norm and DDFs

Figure 7.2   Comparison of the quality of probabilistic constraints after the integration process with Test 2 [51].

b) The effect of choosing coefficient

In this section, therefore, the effect assessment of selecting the coefficients of DDFs is considered on Test 1 with respect to 1-norm. Figure 7.3 presents the influence of coefficient λ in the DDFs A-Div, NA-Div on the quality of the integrating results.

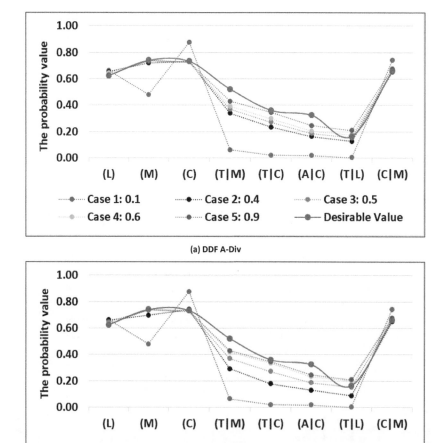

Figure 7.3 Comparison of the integrating results with different coefficients of DDFs A-Div, NA-Div and 1-norm for Test 1 [51].

Figure 7.4 presents the influence of coefficient λ in the DDFs D-Div, CS-Div and R-Div on the quality of the integrating results.

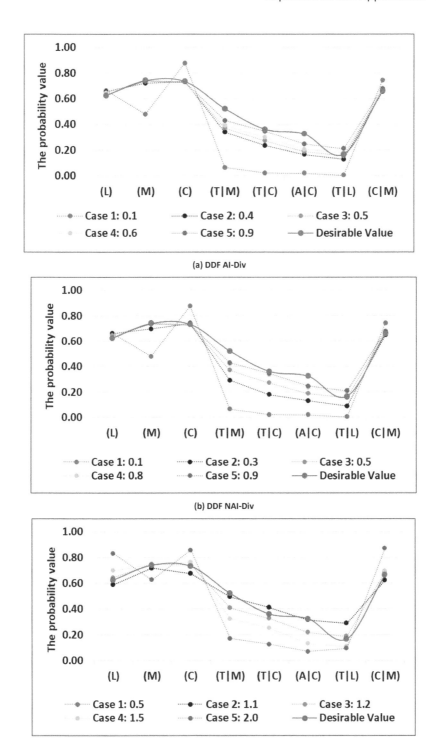

Figure 7.4 Comparison of the integrating results with different coefficients of DDFs AI-Div, NAI-Div and C-Div, and 1-norm for Test 1 [51].

Figure 7.5 presents the influence of coefficient λ in the DDFs **AI**-Div, **NAI**-Div and **C**-Div on the quality of the integrating results.

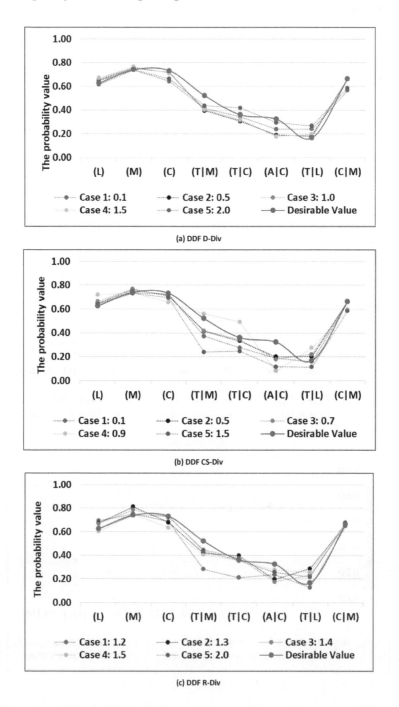

Figure 7.5 Comparison of the integrating results with different coefficients of DDFs D-Div, **CS**-Div and **R**-Div, and 1-norm for Test 1 [51].

The dark red line shows the expected probability values of the constraints that should be reached, while each remaining dotted line indicates the probability values of the constraints gained when λ changes.

The $\lambda \neq 1$ condition applies to the **A**-Div, **NA**-Div, **AI**-Div, **NAI**-Div, **C**-Div, and **D**-Div functions, but the $\frac{-\Pi}{2} \leq \lambda \leq \frac{\Pi}{2}$ condition applies to the **CS**-Div function.

The accuracy of the integrating result can be improved by raising the initial λ, as shown in Figure 7.3a-b. Figure 7.3a shows that the obtained results for $\lambda = 0.1$ are distant from the desired ones, but the obtained results for $\lambda = 0.9$ are close to the desired ones.

Figures 7.5a-c and 7.4a-b demonstrate that the larger λ is, the poorer the quality obtained.

Figure 7.5a shows that the acquired results for $\lambda = 2$ are distant from the desired ones, whereas they are close to the desired ones for $\lambda = 0.5$.

Figure 7.4c shows that the higher the λ, the higher the accuracy of the probability value achieved following the integration process. In the **R**-Div function, however, λ must satisfy the requirement $1 < \lambda \leq 2$.

Overall, the integrating method employing **C**-Div, **D** functions has higher accuracy points than the **AI**-Div, **NAI**-Div, and **CS**-Div functions. For $\lambda = 1.2$, the results are the furthest from the desired ones, while for $\lambda = 2$, the results are the closest. This is consistent with the original idea regarding the appropriate coefficient to use in order to attain the greatest potential result.

c) The computational cost

Figure 7.6 and Figure 7.7 illustrate the computational cost of Test 1 and Test 2, respectively. The number of iterations, the CPU cost of solving the integrating problem, and the total time required to complete the integration process are all taken into account.

- *The CPU cost*: Figure 7.6a and Figure 7.7a shows that the CPU time required to complete integration process on the logarithm DDF group for all norms.

- *The total time to complete the whole integration process*: It depends on three stages:

(i) Restoring consistency.

(ii) Finding the satisfying probability vector.

(iii) Finding probability integrating vector.

Figures 7.6b and 7.7b show the performance of the integration process utilizing DDFs for Test 1 and Test 2, respectively, in terms of three norms. The execution time of Test 1 is in seconds, while Test 2's is in hours.

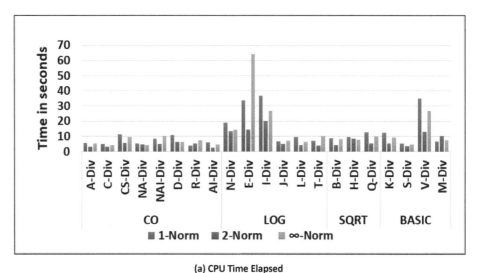

(a) CPU Time Elapsed

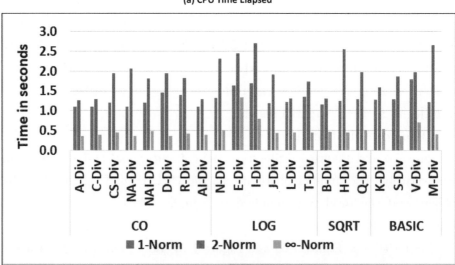

(b) The total execution time

Figure 7.6 The time complexity of the differential algorithm for the particular experiment with Test 1 [51].

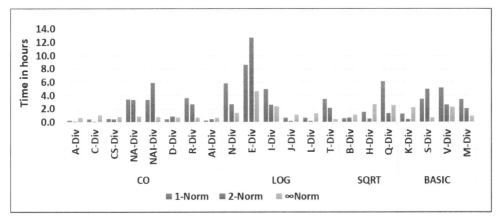

(a) CPU Time Elapsed

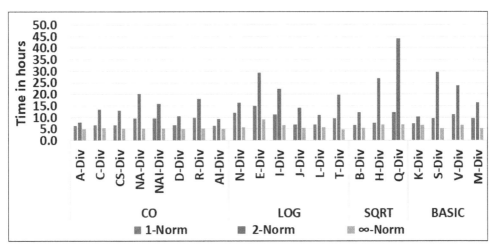

(b) The total execution time

Figure 7.7 The time complexity of the differential algorithm for the particular experiment with Test 2 [51].

In general, during the integration process:

- For most functions used in the integration process, the 2-norm had the greatest cost of norms. This reflects the fact that Algorithm 9 and Algorithm 5 involving the 2-norm have a larger complexity due to the optimization problem being solved with the squared objective function, i.e. $f(\vec{\omega}) = \sqrt{\sum_{i=1}^{h}(a_i^T\vec{\omega})^2}$ (Theorem 5.2) and $g(\vec{x}, \vec{y}) = \sum_{i=1}^{\hbar_E} x_i \cdot log(x_i)$ (Theorem 4.6).

- Both 1-norm and ∞-norm exhibit a large reduction in computational cost because they are done by Algorithm 9 which:

+ The optimization problem with linear objective functions $f(\vec{\omega}, \vec{\lambda}) = \sum_{i=1}^{\hbar_E} \lambda_i$ is solved (Theorem 5.3) with respect to 1-norm.

+ The optimization problem with linear objective functions $f(\vec{\omega}, \lambda) = \lambda$ is solved (Theorem 5.4) with respect to ∞-norm.

The ∞-norm-related algorithms are the most cost-effective because the objective function only has one variable. As a result, the estimate cost of the integrating algorithm implementation in Table 7.1 is consistent with the performance evaluations done on Tests 1 and Test 2. Because the number of optimization variables corresponds to the number of complete conjunctions of E while it is exponential in the size of |E|, the runtime for each integrating algorithm also grows exponentially from Test 1 (time in seconds) in Figure 7.6b to Test 2 (time in hours) in Figure 7.7b.

7.2 APPLICATIONS

7.2.1 Artificial Intelligence and Machine Learning

In this section, we make a survey of typical applications of uncertainty resolution and knowledge integration in AI including Machine Learning, Recommendation Systems, and Group Decision-Making.

7.2.1.1 Machine Learning

In Machine Learning research field, depend on the representation of knowledge, the applications of uncertainty resolution and knowledge integration is classified in three groups, namely Knowledge Graph, Probabilistic Relations, and Logic Rules.

In the group of applications in Knowledge Graph, there are two typical works as follows: In [90], by arguing that text usually contains incomplete and ambiguous information, it causes the difficulties for applying deep learning to solve natural language processing tasks, Bian et al. studies the capacity of leveraging morphological, syntactic, and semantic knowledge is performed to achieve high-quality word embeddings. This knowledge-powered deep learning is proven to enhance the effectiveness of word embedding. Multi-view Factorization AutoEncoder, a framework with network constraints to integrate multi-omic data with domain knowledge is studied by Ma et al. [91]. This framework enables to use deep representation learning to learn both feature embeddings and patient embeddings simultaneously. The feature interaction network and patient view similarity network constraints are integrated into the training objective.

In the group of applications in Probabilistic Relations, the most common works are as follows: Heckerman et al. proposed a Bayesian approach for learning Bayesian networks by integrating prior knowledge and statistical data [92]. Concretely, a methodology for assessing informative priors needed for learning is introduced, the relative posterior probabilities of network structures is computed and search methods for identifying network structures with high posterior probabilities are discussed. In [93], Richardson et al. developed a method for learning the structure of a Bayesian network from multiple experts. Empirical data is then used to refine and parameterize this structure. Montassar et al. [94] proposed a method for learning Causal Bayesian Networks from observational and experimental data by integrating ontological knowledge. In [95], Borboudakis et al. studied the method to integrate the causal relations into a causal model.

Lastly, in the group of applications in Logic Rules. The typical works are as follows: Hu et al. [96, 97] developed a framework that allows to learn knowledge jointly by Deep Neural Networks. By this way, large amounts of fuzzy knowledge can be combined and optimized with a little manual effort. This framework can also be applied to posterior regularization for regulating other statistical models. In [98], a general and principled method to integrate prior knowledge to train deep networks is proposed. In this method, prior knowledge is represented as a collection of first-order logic clauses by Semantic-Based Regularization and each task to be learned corresponds to a predicate in the knowledge base. This task is translated into a set of constraints, and these constraints are integrated into the learning process via back-propagation. Diligenti et al. proposed an approach to learning from constraints which allow to integrate the ability of machine leaning techniques and ability of reasoning with higher-level semantic knowledge [99]. To this end, the learning tasks are modeled in general framework of multi-objective optimization while the constraints translate First Order Logic formulas. Learning the parameters of a Bayesian network from data while taking into account prior knowledge is considered in [100]. The prior knowledge is obtained from domain experts, and parameter learning is a special case of isotonic regression. This study is applied for a small Bayesian network in the medical domain, and it leads to an improved fit of the true distribution with only a small sample of data.

7.2.1.2 Recommendation Systems

The group of projects is in Recommendation Systems with the large range of works such as [101, 102, 103, 104, 105, 106, 107]. The KRAFT project [101] is one of the ealiest project in this direction. This project investigate the way that a Distributed Information System supports to transform and reuse constraints to integrate knowledge and gain extra value. Continuing with KRAFT architecture, Preece et al. proposed an agent-based approach to integate knowledge from various sources by using constraints as common knowledge interchange format [102]. Knowledge in each source is represented by a common constraint language. Integration process is performed via a communicating protocol of agents. Tang et al. [105] proposed a unified topic modeling approach named ArnetMiner. In this model, a topic model is presented simultaneously for modeling papers, authors, and publication venues and the topic model is integrated into a random walk framework. Van Eck and Waltman [107] developed VOSviewer for constructing and viewing bibliometric maps by ontological graphs. Knowledge integration is applied for constructing bibliometric map via similarity matrix and Euclidean distance function. A method to create structured summaries of information based on the metrics of influence, coverage and connectivity for scientific literature is proposed by Shahaf, namely Metro Maps [103]. AKMiner [108] is a system capable of automatically mining useful knowledge to construct conceptual graphs by using Markov Logic Networks (MLN). AceMap [104] is an academic system to analyze the scholarly data by integrating several data mining and network analyse to generate the result in intuitive way. Tao et al. [106] propose a Study-Map Oriented method and RIDP model to study a given paper more efficiently and thoroughly by integrating the designed Reference Injection-based Topic Analysis method and Double-Damping PageRank algorithm in order to mine a Study Map out of massive academic papers. All the above recommendation systems were developed to transform and integrate knowledge from different knowledge sources into a common knowledge base. This knowledge base is ultilized to provide a service to return the results for users. Such services are widely applied for both academic and industrial applications.

7.2.1.3 Group Decision-making

In general, decision theory studies the methods to choose rationally among different alternatives based on their expected utilities [109]. It is a branch of applied probability theory concerned with the theory of making decisions. Group decision-making problem arises when different individuals have different opinions about the probability and utility of alternatives [110, 111, 112, 113]. In order to solve this problem, we have several methods such as these based on communication between actors [114, 115] or taking individual preferences to make group decisions [116, 117]. A typical work for group decision-making via probabilistic belief merging is in [118]. In this work the probabilistic-logic framework [119] and generalized probabilistic entailment [120] are recalled. A decision-theoretical framework containing belief aggregation functions and preference relations is introduced. The relationship between the aggregated group preferences and the preferences of individuals is investigated. The parameters of generalized probabilistic entailment to control the influence of large interest groups are experimented and discussed.

7.2.2 Knowledge Systems

Kim et al. [121] proposed a practical methodology for capturing and representing the knowledge from organizations by using knowledge map to represent knowledge and several procedures to build the knowledge maps. Smart et al. developed a Technical Demonstrator System for the field of situation awareness in the domain of humanitarian operations, especially when these operations in military conflict [122]. In this system, information is collected from heterogeneous data sources, and then it can be interpreted into formal ontological characterizations of the problem domain. Lin and Hsueh [123] proposed a knowledge map management system to facilitate knowledge management by utilizing information retrieval and data mining techniques. This system can create and maintain the dynamic knowledge management of communities of practice on the Internet. Grebla et al. proposed an approach to construct an academic network presented as a learning model to fuse expertise of academic trainers and practitioners by merging knowledge via a mathematical and ontological model [124]. In [125], Cano et al. proposed a new methodology based on Monte Carlo simulations for integrating expert knowledge. It helps to avoid the costly elicitation of these prior distributions. It also requests from the expert only information about those direct probabilistic relationships between variables. Probase is a universal, probabilistic taxonomy which contains large amount of concepts collected from billions of web pages [126]. It utilizes probabilities to model inconsistent, ambiguous and uncertain information thus it has more expressiveness and be potentially applied for text understanding. OLLIE [127] is Open Information Extraction system that allows users to express larger numbers of relation expressions by expanding the syntactic scope of relation phrases, and to expand the Open Information Extraction representation by considering the additional context information such as clausal modifiers and attribution. Knowledge Vault [128] is a Web-scale PKB achieved by integrating the extractions from Web content with prior knowledge collected from existing knowledge repositories. Dong et al. employed several supervised machine learning methods to integrate these knowledge sources. NELL is an agent-based approach to extract information from the web to enrich a structured knowledge base day-by-day [129]. It can also learn to improve the performance of this task over time.

7.2.3 Software Engineering

A software integration event brings multiple software systems together to provide integrated capacity. In [130], Mizell and Malone proposed an approach that uses a software development process simulation model to consider the level of uncertainty arised in an initial estimate. Patnayakuni et al. [131] presents knowledge integration across knowledge boundaries through boundary objects improves information systems development performance. It shows how to enhance the integration of specialized knowledge within and across organizational subunits. It also shows that the positive influence of formal and informal organizational integrative practices on information systems development performance is partially mediated by knowledge integration. Vasantrao et al. [132] discussed how uncertainty creates schedule risk, and then addresses the uncertainty through a framework for model development. This framework includes a five-step modeling process to define the dependencies as the source of the uncertainty. Knowldege collected in the evolution of project will be used to recalibrate in order to decrese the uncertainty. In [133], Steindl and Mottok introduced a metric to calculate the test complexity of a certain integration order. They also proposed an approach for optimizing it with respect to the integration test complexity and the integration test effort using simulated annealing. In [134], a compositional modeling framework for dynamic software product lines is proposed. This framework can support both probabilistic and nondeterministic choices. It also enables timely quantitative analysis and works well for dynamic SPL operational behaviour and family-based analysis based on various quantitative queries. In [135], a methodology for effectively applying knowledge integration to consider both legacy and current data in order to support decision-making for engineering managers. Concretely, it is used to predict schedule delay caused by a software integration error with high accuracy.

7.2.4 Other Applications

Apart from the above groups of applications, uncertainty resolution and knowledge integration is also applied in other applications. In E-Learning, Grebla et al. proposed an approach to construct an academic network presented as a learning model to fuse expertise of academic trainers and practitioners by merging knowledge via a mathematical and ontological model [124].

In BioInformation, EDGAR is a natural language processing system allows to extract information about the relationships between drugs and genes and cancer cure from the biomedical literature [136]. This system was constructed from two resources, namely the MEDLINE database and the Unified Medical Language System.

In agriculture, Nengfu et al. [137] proposed an information access interface by a formal model for knowledge integration from XML data and information access architecture in agriculture to generate agricultural answers from different information sources.

The applications of uncertainty resolution and knowledge integration may also be in the fields of climate change [138, 139], risk assessment [139], and planning decisions in underground mines [140].

CHAPTER 8

Conclusions and open problems

8.1 CONCLUSIONS

The main subject of this book is related to the aspect of knowledge inconsistency resolution and knowledge integration for knowledge bases with probability representations.

In the development history of knowledge-based systems, knowledge integration is a big challenge when knowledge is increasingly expanded and has a probabilistic character. However, to ensure that systems can interact with each other, the probabilistic knowledge bases derived from these systems must be consistent. Therefore, two important problems need to be solved in the process of building a system based on probabily representations are as follows:

- The problem of restoring the consistency of probabilistic knowledge bases.

- The problem of integrating probabilistic knowledge bases.

- The problem of integrating probabilistic knowledge bases. The book presents the extended, more completed, and unified versions of the unpublished PhD thesis [141] of the first author with remaining authors as the co-supervisors, and of our results achieved during the last six years. These results have been included in over 20 scientific papers pub-lished in prestigious international journals (indexed by ISI), or in post-proceedings published by, among others, Springer and IEEEXplore and other international journals and conference proceedings [51, 54, 69, 70, 77, 88, 89, 142]. We acknowledge the permission of Elsevier for including several published tables and figures in this book.

As mentioned in the Preface, the main contributions of this book include the following elements:

- We have surveyed and evaluated existing knowledge integration systems, some methods of dealing with inconsistencies and some methods of integrating knowledge in the form of classical logic, possibilistic logic, probabilistic-logic and probability representation. From there, a general principle diagram of the system integrating probabilistic knowledge has been proposed and compared with existing approaches.

- We have solved the problem of restoring the consistency of the probabilistic knowledge bases by proposing two models of restoring the consistency, namely the models for consistency recovery of standard and non-standard probabilistic knowledge bases, respectively.

- We have solved the problem of integrating probabilistic knowledge bases by proposing two methods, i.e. based on the distances among probabilistic knowledge bases and based on the combination of the probabilistic values of knowledge bases.

We should here mention some limitations of our approach as follows:

- In the schematic diagram of the probabilistic knowledge-based system proposed in this book, we have only built the subsystems including Restoring Consistency and Integrating Probabilistic Knowledge Bases. Building common probabilistic knowledge base from probabilistic knowledge bases has not been built in practice. Reducing probabilistic constraints has only learned a few reduction laws and has not yet conducted experimental settings.

- The process of consistency recovery deals only with the class of inconsistencies of the probabilistic knowledge bases that satisfy a set of expected properties by stating the definitions, related theory without proving these propositions.

- The non-standard consistency recovery model proposed in this book has only proven the construction theorems, proposed processing algorithms and the theorems to evaluate the complexity of the algorithm, but it has not yet implemented real experiments and evaluation.

- The proposed distance-based integrating model of probabilistic knowledge bases only proves the construction theorems; and proposes the processing algorithms and evaluates the complexity of these algorithms. However, the actual experiments and evaluation are omitted.

- The proposed model of integration of probabilistic knowledge bases based on probabilistic values only gives the construction theorems, proposes the processing algorithms and evaluates the complexity of these algorithms. The proof of some theorems and the experiments and evaluation of the theorems are omitted.

- The proposed algorithms to solve the problem of restoring consistency and the problem of integrating probabilistic knowledge bases still have high implementation costs when the input data is large. Concretely, the higher the number of events of the input knowledge base profile, the higher the implementation cost.

8.2   OPEN PROBLEMS

These results do not cover all problems of the subject related to inconsistent knowledge management. It seems that the following problems are very interesting and should be solved in the future:

Some open problems which still require solutions:

- Continuing to study more about probabilistic knowledge base transformation and representation techniques to build and implement subsystems. Build a probabilistic knowledge base from real-world knowledge bases, and about the rules to reduce probabilistic constraints to build and install the subsystem Reduce probabilistic constraints.

- Enhancing the methods to handle the inconsistencies of probabilistic knowledge bases by:

i) Proving theorems about the logical relationship between the class of inconsistencies of probabilistic knowledge bases and a set of expected properties;

ii) Proposing and installing algorithms to calculate the baseline measures and using them in the model to restore the consistency of the probabilistic knowledge bases;

iii) Proceeding to install the non-standard consistency recovery algorithm on the actual experimental sets.

- Perfecting methods of integrating probabilistic knowledge bases. Installing Hull Algorithms and Omma Algorithms on real experimental sets. From there evaluate the efficiency of the algorithm as well as the results obtained.

- Improving the algorithms towards more efficiency in computing performance as well as build a standard dataset for further research and experiments.

Bibliography

[1] Ngoc Thanh Nguyen. *Advanced Methods for Inconsistent Knowledge Management*. Advanced Information and Knowledge Processing. Springer, 2008.

[2] Isabelle Bloch, Anthony Hunter, Alain Appriou, André Ayoun, Salem Benferhat, Philippe Besnard, Laurence Cholvy, and Roger M. Cooke. Fusion: General concepts and characteristics. *Int. J. Intell. Syst.*, 16(10):1107–1134, 2001.

[3] Carlos E. Alchourrón, Peter Gärdenfors, and David Makinson. On the logic of theory change: Partial meet contraction and revision functions. *J. Symb. Log.*, 50(2):510–530, 1985.

[4] Raymond Reiter. A theory of diagnosis from first principles. *J. Artif. Intell.*, 32(1):57–95, 1987.

[5] Sven Ove Hansson. *A textbook of belief dynamics*, volume 11/1 of *Applied Logic Series*. Springer Netherlands, 1999.

[6] Marc Finthammer, Gabriele Kern-Isberner, and Manuela Ritterskamp. Resolving inconsistencies in probabilistic knowledge bases. In *Proceedings of 30th Annual German Conference on Advances in Artificial Intelligence*, pages 114–128, 2007.

[7] Anthony Hunter and Sébastien Konieczny. Shapley inconsistency values. In *Proceedings of the 10th International Conference on Principles of Knowledge Representation and Reasoning*, pages 249–259, 2006.

[8] Anthony Hunter and Sébastien Konieczny. Measuring inconsistency through minimal inconsistent sets. In *Proceedings of the 11th International Conference on Principles of Knowledge Representation and Reasoning*, pages 358–366, 2008.

[9] Anthony Hunter and Sébastien Konieczny. On the measure of conflicts: Shapley inconsistency values. *J. Artif. Intell.*, 174(14):1007–1026, 2010.

[10] Mark Liffiton and Karem Sakallah. On finding all minimally unsatisfiable sub-formulas. In *Proceedings of the 8th International Conference on Theory and Applications of Satisfiability Testing*, pages 173–186, 2005.

[11] Wilhelm Rödder and Longgui Xu. Elimination of inconsistent knowledge in the probabilistic expert system-shell spirit (in German). In *Proceedings of Symposium Operations Research 2000*, pages 260–265, 2001.

[12] Gabriele Kern-Isberner and Wilhelm Rödder. Belief revision and information fusion in a probabilistic environment. In *Proceedings of the 16th International Florida Artificial Intelligence Research Society Conference*, pages 506–510, 2003.

[13] Salem Benferhat, Didier Dubois, Souhila Kaci, and Henri Prade. Possibilistic merging and distance-based fusion of propositional information. *J. Ann. Math. Artif. Intell.*, 34(1-3):217–252, 2002.

[14] Sébastien Konieczny, Jérôme Lang, and Pierre Marquis. Da^2 merging operators. *J. Artif. Intell.*, 157(1-2):49–79, 2004.

[15] Guilin Qi, Weiru Liu, and David A. Bell. Merging stratified knowledge bases under constraints. In *Proceedings of the 21st National Conference on Artificial Intelligence and the Eighteenth Innovative Applications of Artificial Intelligence Conference*, pages 281–286, 2006.

[16] Guilin Qi, Jianfeng Du, Weiru Liu, and David A. Bell. Merging knowledge bases in possibilistic logic by lexicographic aggregation. *j. CoRR*, abs/1203.3508, 2012.

[17] Sébastien Konieczny and Ramón Pino Pérez. Logic based merging. *J. Philosophical Log.*, 40(2):239–270, 2011.

[18] Adrian Haret. Logic-based merging in fragments of classical logic with inputs from social choice theory. In *Proceedings of the 5th International Conference on Algorithmic Decision Theory*, pages 374–378, 2017.

[19] Sébastien Konieczny. Belief base merging as a game. *J. Applied Non-Classical Log.*, 14(3):275–294, 2004.

[20] Dongmo Zhang. A logic-based axiomatic model of bargaining. *J. Artif. Intell.*, 174(16-17):1307–1322, 2010.

[21] Trong Hieu Tran, Thi Hong Khanh Nguyen, Quang-Thuy Ha, and Ngoc Trinh Vu. Argumentation framework for merging stratified belief bases. In *Proceedings of the 8th Asian Conference on Intelligent Information and Database Systems*, pages 43–53, 2016.

[22] Thi Hong Khanh Nguyen, Trong Hieu Tran, Tran Van Nguyen, and Thi Thanh Luu Le. Merging possibilistic belief bases by argumentation. In *Proceedings of the 9th Asian Conference on Intelligent Information and Database Systems*, pages 24–34, 2017.

[23] Nico Potyka, Erman Acar, Matthias Thimm, and Heiner Stuckenschmidt. Group decision making via probabilistic belief merging. In *Proceedings of the 25th International Joint Conference on Artificial Intelligence*, pages 3623–3629, 2016.

[24] Jiří Vomlel. *Methods of Probabilistic Knowledge Integration*. PhD thesis, Czech Technical University, Prague, 1999.

[25] Jirí Vomlel. Integrating inconsistent data in a probabilistic model. *J. Appl. Non Class. Logi.*, 14(3):367–386, 2004.

[26] Shenyong Zhang, Yun Peng, and Xiaopu Wang. An efficient method for probabilistic knowledge integration. In *Proceedings of the 20th IEEE International Conference on Tools with Artificial Intelligence*, pages 179–182, 2008.

[27] George Wilmers. The social entropy process: Axiomatising the aggregation of probabilistic beliefs. In *Probability, Uncertainty and Rationality*, pages 1–19. The University of Manchester, 2010.

[28] Martin Adamcík and George Wilmers. The irrelevant information principle for collective probabilistic reasoning. *J. Kybernetika*, 50(2):175–188, 2014.

[29] Martin Adamcík. *Collective Reasoning under Uncertainty and Inconsistency*. PhD thesis, University of Manchester, UK., 2014.

[30] George Wilmers. A foundational approach to generalising the maximum entropy inference process to the multi-agent context. *J. Entropy*, 17(2):594–645, 2015.

[31] William B Levy and Hakan Deliç. Maximum entropy aggregation of individual opinions. *J. IEEE Transactions on Systems, Man, and Cyber.*, 24(4):606–613, April 1994.

[32] Gabriele Kern-Isberner and Wilhelm Rödder. Belief revision and information fusion on optimum entropy. *Int. J. Intell. Syst.*, 19(9):837–857, 2004.

[33] Gabriele Kern-Isberner, Marco Wilhelm, and Christoph Beierle. Probabilistic knowledge representation using the principle of maximum entropy and gröbner basis theory. *J. Ann. Math. Artif. Intell.*, 79(1-3):163–179, 2017.

[34] Lionel Daniel. *Paraconsistent Probabilistic Reasoning: applied to scenario recognition and voting theory.* PhD thesis, Automatic. École Nationale Supérieure des Mines de Paris. English., 2010.

[35] Frank van Harmelen, Vladimir Lifschitz, and Bruce Porter, editors. *Handbook of Knowledge Representation*, volume 3 of *Foundations of Artificial Intelligence*. Elsevier, 2008.

[36] Gabriele Kern-Isberner. *Conditionals in Nonmonotonic Reasoning and Belief Revision - Considering Conditionals as Agents*, volume 2087 of *Lecture Notes in Computer Science*. Springer, 2001.

[37] Ronald Walpole, Raymond Myers, Sharon Myers, and Keying Ye. *Probability and Statistics for Engineers and Scientists*. Prentice Hall, 2012.

[38] Matthias Thimm. *Probabilistic Reasoning with Incomplete and Inconsistent Belief.* PhD thesis, Technische Universität Dortmund, Germany, 2011.

[39] Nico Potyka. Linear programs for measuring inconsistency in probabilistic logics. In *Proceedings of the 14th International Conference on Principles of Knowledge Representation and Reasoning*, pages 568–577, 2014.

[40] Nico Potyka. *Solving Reasoning Problems for Probabilistic Conditional Logics with Consistent and Inconsistent Information.* PhD thesis, FernUniversitat, Hagen, 2016.

[41] Enrique F. Castillo, José Manuel Gutiérrez, and Ali S. Hadi. *Expert Systems and Probabilistic Network Models*. Monographs in Computer Science. Springer, 1997.

[42] Muhammad Saiful Islam, Madhav Prasad Nepal, Ronald Martin Skitmore, and Golam Kabir. A knowledge-based expert system to assess power plant project cost overrun risks. *J. Expert Syst. Appl.*, 136:12–32, 2019.

[43] Yang Chen and Daisy Zhe Wang. Knowledge expansion over probabilistic knowledge bases. In *International Conference on Management of Data, SIGMOD 2014, Snowbird, Utah, June 22-27, 2014*, pages 649–660, 2014.

[44] Gaëtan Blondet, Julien Le Duigou, and Nassim Boudaoud. A knowledge-based system for numerical design of experiments processes in mechanical engineering. *J. Expert Syst. Appl.*, 122:289–302, 2019.

[45] Pavel Klinov, Bijan Parsia, and David Picado-Muiño. The consistency of the medical expert system CADIAG-2: A probabilistic approach. *J. Information Technology Rese.*, 4(1):1–20, 2011.

[46] David Poole. Average-case analysis of a search algorithm for estimating prior and posterior probabilities in bayesian networks with extreme probabilities. In *Proceedings of the 13th International Joint Conference on Artificial Intelligence*, pages 606–612, 1993.

[47] Adnan Darwiche. Conditioning methods for exact and approximate inference in causal networks. In *Proceedings of the 11th Conference on Uncertainty in Artificial Intelligence*, pages 99–107. Morgan Kauffman, 1995.

[48] Francisco Díez. Local conditioning in bayesian networks. *J. Artif. Intell.*, 87:1–20, 11 1996.

[49] Xiaojun Chen, Shengbin Jia, and Yang Xiang. A review: Knowledge reasoning over knowledge graph. *J. Expert Syst. Appl.*, (141):112948, 2020.

[50] Wei Yang, Chaofan Fu, Xiaoguang Yan, and Zhuoning Chen. A knowledge-based system for quality analysis in model-based design. *J. Intell. Manuf.*, 31(6):1579–1606, 2020.

[51] Van Tham Nguyen, Trong Hieu Tran, and Ngoc Thanh Nguyen. A model for building probabilistic knowledge-based systems using divergence distances. *J. Expert Syst. Appl.*, 174:114494, 2021.

[52] Christopher D. Manning and Hinrich Schütze. *Foundations of statistical natural language processing.* MIT Press, 2001.

[53] Matthias Thimm. Inconsistency measures for probabilistic logics. *J. Artif. Intell.*, 197:1–24, 2013.

[54] Van Tham Nguyen, Ngoc Thanh Nguyen, Trong Hieu Tran, and Kieu Loan Nguyen Do. Method for restoring consistency in probabilistic knowledge bases. *J. Cybernetics and Sys.*, 49(0):1–22, 2018.

[55] Aman Ullah. Entropy, divergence and distance measures with econometric applications. *J. Stati. Planning and Infe.*, 49(1):137–162, 1996. Econometric Methodology, Part I.

[56] Stephen Boyd and Lieven Vandenberghe, editors. *Convex Optimization*. New York: Cambridge, 2004.

[57] Glauber De Bona. *Measuring Inconsistency in Probabilistic Knowledge Bases*. PhD thesis, University of Sao Paulo, 2016.

[58] Dragan Doder, Miodrag Raskovic, Zoran Markovic, and Zoran Ognjanovic. Measures of inconsistency and defaults. *Int. J. Approx. Reasoning*, 51(7):832–845, 2010.

[59] John Grant and Anthony Hunter. Measuring consistency gain and information loss in stepwise inconsistency resolution. In *Proceedings of the 11th European Conference on Symbolic and Quantitative Approaches to Reasoning with Uncertainty*, pages 362–373, 2011.

[60] John Grant and Anthony Hunter. Distance-based measures of inconsistency. volume 7958 of *Lecture Notes in Computer Science*, pages 230–241. Springer, 2013.

[61] Matthias Thimm. On the compliance of rationality postulates for inconsistency measures: A more or less complete picture. *J. Kidney Inte.*, 31(1):31–39, 2017.

[62] Hans van Maaren and Siert Wieringa. Finding guaranteed muses fast. In *Proceedings of the 11th International Conference on Theory and Applications of Satisfiability Testing*, pages 291–304, 2008.

[63] João P. Marques Silva. Minimal unsatisfiability: Models, algorithms and applications (invited paper). In *40th IEEE International Symposium on Multiple-Valued Logic, ISMVL 2010, Barcelona, Spain, 26-28 May 2010*, pages 9–14, 2010.

[64] Sébastien Konieczny and Stéphanie Roussel. A reasoning platform based on the MI shapley inconsistency value. In *Proceedings of the 12th European Conference*

on Symbolic and Quantitative Approaches to Reasoning with Uncertainty, pages 315–327, 2013.

[65] Matthias Thimm. Measuring inconsistency in probabilistic knowledge bases. In *Proceedings of the 25th Conference on Uncertainty in Artificial Intelligence, Montreal*, pages 530–537, 2009.

[66] Nico Potyka and Matthias Thimm. Probabilistic reasoning with inconsistent beliefs using inconsistency measures. In *Proceedings of the 24th International Joint Conference on Artificial Intelligence*, pages 3156–3163, 2015.

[67] David Picado-Muiño. Measuring and repairing inconsistency in probabilistic knowledge bases. *Int. J. Approx. Reasoning*, 52(6):828–840, 2011.

[68] Dimitri P. Bertsekas, Angelia Nedic, and Asuman E. Ozdaglar, editors. *Convex Analysis and Optimization*. Athena Scientific, 2003.

[69] Van Tham Nguyen and Trong Hieu Tran. Inconsistency measures for probabilistic knowledge bases. In *9th International Conference on Knowledge and Systems Engineering, KSE 2017, Hue, Vietnam, October 19-21, 2017*, pages 148–153. IEEE, 2017.

[70] Van Tham Nguyen and Trong Hieu Tran. Solving inconsistencies in probabilistic knowledge bases via inconsistency measures. In *Intelligent Information and Database Systems - 10th Asian Conference, ACIIDS 2018, Dong Hoi City, Vietnam, March 19-21, 2018, Proceedings, Part I*, volume 10751 of *Lecture Notes in Computer Science*, pages 3–14. Springer, 2018.

[71] Jiri Matousek and Bernd Gärtner. Understanding and using linear programming. In *Universitext*, pages 1–222. Springer-Verlag Berlin Heidelberg, 2007.

[72] Radosław Katarzyniak and Ngoc Thanh Nguyen. Reconciling inconsistent profiles of agent's knowledge states in distributed multiagent systems using consensus methods. *Int. J. SySc*, 26:93–119, 2000.

[73] Ngoc Thanh Nguyen. Consensus system for solving conflicts in distributed systems. *J. Information Scie.*, 147(1-4):91–122, 2002.

[74] Grzegorz Kolaczek, Agnieszka Pieczynska-Kuchtiak, Krzysztof Juszczyszyn, Adam Grzech, Radoslaw Katarzyniak, and Ngoc Thanh Nguyen. A mobile agent approach to intrusion detection in network systems. In *Proceedings of*

the *9th International Conference on Knowledge-Based Intelligent Information and Engineering Systems*, pages 514–519, 2005.

[75] Ngoc Thanh Nguyen. Processing inconsistency of knowledge on semantic level. *J. Universal Computer Scie.*, 11(2):285–302, 2005.

[76] Nico Potyka and Matthias Thimm. Consolidation of probabilistic knowledge bases by inconsistency minimization. In *Proceedings of the 21st European Conference on Artificial Intelligence*, pages 729–734, 2014.

[77] Van Tham Nguyen, Trong Hieu Tran, Ngoc Thanh Nguyen, and Do Kieu Loan Nguyen. Resolving inconsistencies in probabilistic knowledge bases by quantitative modification. In *13th International Conference on Knowledge and Systems Engineering, KSE 2021, Bangkok, Thailand, November 10-12, 2021*, pages 1–4. IEEE, 2021.

[78] Patricia Everaere, Sébastien Konieczny, and Pierre Marquis. Disjunctive merging: Quota and gmin merging operators. *j. Artif. Intell.*, 174(12-13):824–849, 2010.

[79] Guilin Qi, Weiru Liu, and David H. Glass. A split-combination method for merging inconsistent possibilistic knowledge bases. In *Proceedings of the 9th International Conference on Principles of Knowledge Representation and Reasoning*, pages 348–356, 2004.

[80] Guilin Qi, Weiru Liu, and David A. Bell. A revision-based approach to resolving conflicting information. In *Proceedings of the 21st Conference in Uncertainty in Artificial Intelligence*, pages 477–484, 2005.

[81] Guilin Qi, Weiru Liu, David H. Glass, and David A. Bell. A split-combination approach to merging knowledge bases in possibilistic logic. *J. Ann. Math. Artif. Intell.*, 48(1-2):45–84, 2006.

[82] Arak Mathai and Pushpa Rathie. Basic concepts in information theory and statistics. *J. SERBIULA (sistema Librum 2.0)*, 1975.

[83] Shinto Eguchi. A differential geometric approach to statistical inference on the basis of contrast functionals. *J. Hiroshima Math.*, 15(2):341–391, 1985.

[84] Jongkyeong Chung, Palaniappan Kannappan, Che Tat Ng, and Prasanna Sahoo. Measures of distance between probability distributions. *J. Math. Analysis Applic.*, 138:280–292, 02 1989.

[85] Jianhua Lin. Divergence measures based on the shannon entropy. *J. IEEE Trans. Information Theory*, 37(1):145–151, 1991.

[86] Inder Jeet Taneja. *Generalized Information Measures and Their Applications.* Universidade Federal de Santa Catarina, 2001.

[87] Gabriel Martos Venturini. *Statistical Distances and Probability Metrics for Multivariate Data, Ensembles and Probability Distributions.* PhD thesis, UNIVERSIDAD CARLOS III DE MADRI, Spain, 2015.

[88] Van Nguyen, Ngoc Thanh Nguyen, and Trong Hieu Tran. A distance-based approach for merging probabilistic knowledge bases. *J. Intelligent and Fuzzy Sys.*, pages 1–14, 07 2019.

[89] Van Tham Nguyen, Ngoc Thanh Nguyen, and Trong Hieu Tran. Algorithms for merging probabilistic knowledge bases. In *Intelligent Information and Database Systems - 11th Asian Conference, ACIIDS 2019, Yogyakarta, Indonesia, April 8-11, 2019, Proceedings, Part I*, pages 3–15, 2019.

[90] Jiang Bian, Bin Gao, and Tie-Yan Liu. Knowledge-powered deep learning for word embedding. In *Proceedings of the 2014th European Conference on Machine Learning and Knowledge Discovery in Databases, Volume Part I*, ECMLP-KDD'14, page 132–148, Berlin, Heidelberg, 2014. Springer-Verlag.

[91] Tianle Ma and Aidong Zhang. Multi-view factorization autoencoder with network constraints for multi-omic integrative analysis. In *2018 IEEE International Conference on Bioinformatics and Biomedicine (BIBM)*, pages 702–707, 2018.

[92] David Heckerman, Dan Geiger, and David M. Chickering. Learning bayesian networks: The combination of knowledge and statistical data. In Ramon Lopez de Mantaras and David Poole, editors, *Uncertainty Proceedings 1994*, pages 293–301. Morgan Kaufmann, San Francisco (CA), 1994.

[93] Matthew Richardson and Pedro Domingos. Learning with knowledge from multiple experts. In *Proceedings of the Twentieth International Conference on International Conference on Machine Learning*, ICML'03, page 624–631. AAAI Press, 2003.

[94] Montassar Ben Messaoud, Philippe Leray, and Nahla Ben Amor. Integrating ontological knowledge for iterative causal discovery and visualization. In

Claudio Sossai and Gaetano Chemello, editors, *Symbolic and Quantitative Approaches to Reasoning with Uncertainty*, pages 168–179, Berlin, Heidelberg, 2009. Springer Berlin Heidelberg.

[95] Giorgos Borboudakis and Ioannis Tsamardinos. Incorporating causal prior knowledge as path-constraints in bayesian networks and maximal ancestral graphs. In *Proceedings of the 29th International Conference on Machine Learning, ICML 2012, Edinburgh, Scotland, UK, June 26 - July 1, 2012*. icml.cc / Omnipress, 2012.

[96] Zhiting Hu, Zichao Yang, Ruslan Salakhutdinov, and Eric Xing. Deep neural networks with massive learned knowledge. In *Proceedings of the 2016 Conference on Empirical Methods in Natural Language Processing*, pages 1670–1679, Austin, Texas, November 2016. Association for Computational Linguistics.

[97] Zhiting Hu, Xuezhe Ma, Zhengzhong Liu, Eduard Hovy, and Eric Xing. Harnessing deep neural networks with logic rules. In *Proceedings of the 54th Annual Meeting of the Association for Computational Linguistics (Volume 1: Long Papers)*, pages 2410–2420, Berlin, Germany, August 2016. Association for Computational Linguistics.

[98] Michelangelo Diligenti, Soumali Roychowdhury, and Marco Gori. Integrating prior knowledge into deep learning. In *2017 16th IEEE International Conference on Machine Learning and Applications (ICMLA)*, pages 920–923, 2017.

[99] Michelangelo Diligenti, Marco Gori, and Claudio Saccà. Semantic-based regularization for learning and inference. *J. Arti Intelligence*, 244:143–165, 2017. Combining Constraint Solving with Mining and Learning.

[100] Ad Feelders and Linda C. van der Gaag. Learning bayesian network parameters under order constraints. *Int. J. Approximate Reas.*, 42(1):37–53, 2006. PGM'04.

[101] Peter Gray, Alun Preece, Nicholas John Fiddian, Alex Gray, Trevor Bench-Capon, Mike Shave, Martin Beer, Zhan Cui, Brillid Diaz, Suzanne Embury, Kit ying Hui, Andrew Jones, Dean Jones, Ken Lunn, Pepijn Visser, and et al. Kraft: Knowledge fusion from distributed databases and knowledge bases. *International Conference on Database and Expert Systems Applications – DEXA*, 04 1997.

[102] Alun Preece, Kitying Ying Hui, Alex Gray, Paul Martin, Trevor Bench-Capon, David Jones, and Zhanfeng Cui. The kraft architecture for knowledge fusion and transformation. *J. Knowledge-Based Sys.*, 13(2):113–120, 2000.

[103] Dafna Shahaf, Carlos Guestrin, and Eric Horvitz. Metro maps of science. In *Proceedings of the 18th ACM SIGKDD International Conference on Knowledge Discovery and Data Mining*, KDD '12, page 1122–1130, New York, NY, 2012. Association for Computing Machinery.

[104] Zhaowei Tan, Changfeng Liu, Yuning Mao, Yunqi Guo, Jiaming Shen, and Xinbing Wang. Acemap: A novel approach towards displaying relationship among academic literatures. In *Proceedings of the 25th International Conference Companion on World Wide Web*, WWW '16 Companion, page 437–442, Republic and Canton of Geneva, CHE, 2016. International World Wide Web Conferences Steering Committee.

[105] Jie Tang, Ruoming Jin, and Jing Zhang. A topic modeling approach and its integration into the random walk framework for academic search. In *2008 Eighth IEEE International Conference on Data Mining*, pages 1055–1060, 2008.

[106] Shibo Tao, Xiaorong Wang, Weijing Huang, Wei Chen, Tengjiao Wang, and Kai Lei. From citation network to study map: A novel model to reorganize academic literatures. In *Proceedings of the 26th International Conference on World Wide Web Companion*, WWW '17 Companion, page 1225–1232, Republic and Canton of Geneva, CHE, 2017. International World Wide Web Conferences Steering Committee.

[107] Nees Jan van Eck and Ludo Waltman. Software survey: Vosviewer, a computer program for bibliometric mapping. *J. Scientometrics*, 84(2):523–538, 2010.

[108] Shanshan Huang and Xiaojun Wan. Akminer: Domain-specific knowledge graph mining from academic literatures. In Xuemin Lin, Yannis Manolopoulos, Divesh Srivastava, and Guangyan Huang, editors, *Web Information Systems Engineering – WISE 2013*, pages 241–255, Berlin, Heidelberg, 2013. Springer Berlin Heidelberg.

[109] Simon French, editor. *Decision Theory: An Introduction to the Mathematics of Rationality*. Halsted Press, 1986.

[110] Christopher P. Chambers and Takashi Hayashi. Preference aggregation under uncertainty: Savage vs. Pareto. *J. Games and Economic Beha.*, 54(2):430–440, February 2006.

[111] Hervé Crès, Itzhak Gilboa, and Nicolas Vieille. Aggregation of multiple prior opinions. *J. Economic Theo.*, 146(6):2563–2582, 2011.

[112] Itzhak Gilboa, Dov Samet, and David Schmeidler. Utilitarian aggregation of beliefs and tastes. *J. Political Eco.*, 112(4):932–938, 2004.

[113] Klaus Nehring. The impossibility of a paretian rational: A bayesian perspective. *J. Economics Let.*, 96(1):45–50, 2007.

[114] Pietro Panzarasa, Nicholas R. Jennings, and Timothy J. Norman. Formalizing Collaborative Decision-making and Practical Reasoning in Multi-agent Systems. *J. Logic and Compu.*, 12(1):55–117, 02 2002.

[115] M Wooldridge and NR Jennings. The cooperative problem-solving process. *J. Logic and Compu.*, 9(4):563–592, 08 1999.

[116] Hervé Moulin. *Handbook of Computational Social Choice*. Cambridge University Press, 2016.

[117] Yoav Shoham and Kevin Leyton-Brown. *Multiagent Systems: Algorithmic, Game-Theoretic, and Logical Foundations*. Cambridge University Press, 2008.

[118] Nico Potyka, Erman Acar, Matthias Thimm, and Heiner Stuckenschmidt. Group decision making via probabilistic belief merging. In *Proceedings of the Twenty-Fifth International Joint Conference on Artificial Intelligence*, IJCAI'16, page 3623–3629. AAAI Press, 2016.

[119] Thomas Lukasiewicz. Probabilistic deduction with conditional constraints over basic events. *J. Artif. Int. Res.*, 10(1):199–241, apr 1999.

[120] Nico Potyka and Matthias Thimm. Consolidation of probabilistic knowledge bases by inconsistency minimization. In *Proceedings of the Twenty-First European Conference on Artificial Intelligence*, ECAI'14, page 729–734, NLD, 2014. IOS Press.

[121] Suyeon Kim, Euiho Suh, and Hyunseok Hwang. Building the knowledge map: an industrial case study. *J. Knowledge Manage.*, 7(2):34–45, 2003.

[122] Paul Smart, Nigel Shadbolt, Leslie Carr, and Mc Schraefel. Knowledge-based information fusion for improved situational awareness. In *2005 7th International Conference on Information Fusion*, volume 2, pages 1018–1024, 2005.

[123] Fu ren Lin and Chih ming Hsueh. Knowledge map creation and maintenance for virtual communities of practice. *Information Processing – Management*, 42(2):551–568, 2006.

[124] Horea Adrian Grebla, Calin Ovidiu Cenan, and Liliana Stanca. Knowledge fusion in academic networks. *BRAIN. Broad Research in Artificial Intelligence and Neuroscience*, 1(2):pp. 111–118, Apr. 2010.

[125] Andrés Cano, Andrés Masegosa, and Serafín Moral. A method for integrating expert knowledge when learning bayesian networks from data. *j. IEEE transactions on systems, man, and cyber.. Part B, Cybernetics : a publication of the IEEE Systems, Man, and Cybernetics Society*, 41:1382–94, 06 2011.

[126] Wentao Wu, Hongsong Li, Haixun Wang, and Kenny Zhu. Probase: A probabilistic taxonomy for text understanding. In *Proceedings of the 2012 ACM SIGMOD International Conference on Management of Data*, SIGMOD '12, page 481–492, New York, NY, 2012. Association for Computing Machinery.

[127] Mausam, Michael Schmitz, Robert Bart, Stephen Soderland, and Oren Etzioni. Open language learning for information extraction. In *Proceedings of the 2012 Joint Conference on Empirical Methods in Natural Language Processing and Computational Natural Language Learning*, EMNLP-CoNLL '12, page 523–534, USA, 2012. Association for Computational Linguistics.

[128] Xin Dong, Evgeniy Gabrilovich, Geremy Heitz, Wilko Horn, Ni Lao, Kevin Murphy, Thomas Strohmann, Shaohua Sun, and Wei Zhang. Knowledge vault: A web-scale approach to probabilistic knowledge fusion. In *Proceedings of the 20th ACM SIGKDD International Conference on Knowledge Discovery and Data Mining*, KDD '14, page 601–610, New York, NY, 2014. Association for Computing Machinery.

[129] Andrew Carlson, Justin Betteridge, Bryan Kisiel, Burr Settles, Estevam R. Hruschka, and Tom M. Mitchell. Toward an architecture for never-ending language learning. In *Proceedings of the Twenty-Fourth AAAI Conference on Artificial Intelligence*, AAAI'10, page 1306–1313. AAAI Press, 2010.

[130] Carolyn Mizell and Linda Malone. A project management approach to using simulation for cost estimation on large, complex software development projects. *J. Engineering Manage.*, 19(4):28–34, 2007.

[131] Ravi Patnayakuni, Arun Rai, and Amrit Tiwana. Systems development process improvement: A knowledge integration perspective. *IEEE Transactions on Engineering Management*, 54(2):286–300, 2007.

[132] Kardile Vilas Vasantrao. Enhance accuracy in software cost and schedule estimation by using "Uncertainty Analysis and Assessment" in the system modeling process. *Int. J. Research and Innovation in Computer Eng.*, 1:6–18, 2011.

[133] Michael Steindl and Juergen Mottok. Optimizing software integration by considering integration test complexity and test effort. In *Proceedings of the 10th International Workshop on Intelligent Solutions in Embedded Systems*, pages 63–68, 2012.

[134] Clemens Dubslaff, Sascha Klüppelholz, and Christel Baier. Probabilistic model checking for energy analysis in software product lines. In *Proceedings of the 13th International Conference on Modularity*, MODULARITY '14, page 169–180, New York, NY, 2014. Association for Computing Machinery.

[135] Kelly D. Alexander, Thomas A. Mazzuchi, and Shahram Sarkani. Knowledge integration for predicting schedule delays in software integration. *J. Engineering Manage.*, 29(4):223–234, 2017.

[136] Thomas Rindflesch, Lorraine Tanabe, John Weinstein, and Lawrence Hunter. Edgar: extraction of drugs, genes and relations from the biomedical literature. *Pacific Symposium on Biocomputing. Pacific Symposium on Biocomputing*, (10902199):517–528, 2000.

[137] Xie Nengfu, Wang Wensheng, Yang Xiaorong, and Jiang Lihua. Rule-based agricultural knowledge fusion in web information integration. *Sensor Letters*, 10(1-2):635–638, 2012.

[138] Joeri Rogelj, David McCollum, Andy Reisinger, Malte Meinshausen, and Keywan Riahi. Integrating uncertainties for climate change mitigation. In *EGU General Assembly Conference Abstracts*, EGU General Assembly Conference Abstracts, pages EGU2013–1848, April 2013.

[139] Howard Kunreuther, Shreekant Gupta, Valentina Bosetti, Roger Cooke, Varun Dutt, Minh Ha-Duong, Hermann Held, Juan Llanes-Regueiro, Anthony Patt, Ekundayo Shittu, and Elke Weber. Chapter 2 - integrated risk and uncertainty assessment of climate change response policies. In *Climate Change 2014: Mitigation of Climate Change. IPCC Working Group III Contribution to AR5.* Cambridge University Press, November 2014.

[140] João Ricardo Gonçalves Neves, Cristina da Paixão Araújo, and Amílcar Soares. Uncertainty integration in dynamic mining reserves. *J. Mathematical Geos.*, 53:737–755, 2020.

[141] Van Tham Nguyen. *Restoring the consistency and merging knowledge bases using the probability model.* PhD thesis, VNU University of Engineering and Technology, Vietnam National University, Hanoi, Vietnam, 2021.

[142] Van Tham Nguyen, Ngoc Thanh Nguyen, and Trong Hieu Tran. Framework for merging probabilistic knowledge bases. In *Computational Collective Intelligence - 10th International Conference, ICCCI 2018, Bristol, UK, September 5-7, 2018, Proceedings, Part I*, volume 11055 of *Lecture Notes in Computer Science*, pages 31–42. Springer, 2018.

Index